New Interventionist Just War Theory

IIIII I IIIIIIIIIIIIIIIII III III
I0131711

This book offers a systematic critique of recent interventionist just war theories, which have made the recourse to force easier to justify.

The work argues that these theories, including neo-traditionalist prerogatives to national leaders and a cosmopolitan human rights paradigm, offer criteria for war that are insufficient in principle and dangerous in practice. Drawing on a plurality of moral considerations, the book recommends a modified legalist national defense paradigm, which includes an atrocity threshold for humanitarian intervention and a legitimate authorization requirement. The plausibility of this restrictive framework is applied to case studies, including the long wars in Iraq and Afghanistan, ongoing targeted killing, and possible interventions in Syria and elsewhere. Various arguments which seek to loosen the criteria for war are also systematically analyzed and criticized.

This book will be of much interest to students of just war theory, military history, ethics, political philosophy, and international relations.

Jordy Rocheleau is Professor of Philosophy at Austin Peay State University, USA. He is co-author of *Rights and Wrongs in the College Classroom* (2007).

Routledge Studies in Intervention and Statebuilding

The series publishes monographs and edited collections analysing a wide range of policy interventions associated with statebuilding. It asks broader questions about the dynamics, purposes and goals of this interventionist framework and assesses the impact of externally-guided policy-making.

Series Editors: Aidan Hehir, *University of Westminster, UK*, Pol Bargues, *CIDOB (Barcelona Centre for International Affairs), Spain*, and Vjosa Musliu, *Vrije Universiteit Brussel (VUB), Belgium*

For more information about this series, please visit: https://www.routledge.com/Routledge-Studies-in-Intervention-and-Statebuilding/book-series/RSIS

New Interventionist Just War Theory

A Critique

Jordy Rocheleau

Routledge
Taylor & Francis Group

LONDON AND NEW YORK

First published 2022
by Routledge
2 Park Square, Milton Park, Abingdon, Oxon OX14 4RN

and by Routledge
605 Third Avenue, New York, NY 10158

Routledge is an imprint of the Taylor & Francis Group, an informa business

© 2022 Jordy Rocheleau

The right of Jordy Rocheleau to be identified as author of this work has been asserted by him in accordance with sections 77 and 78 of the Copyright, Designs and Patents Act 1988.

All rights reserved. No part of this book may be reprinted or reproduced or utilised in any form or by any electronic, mechanical, or other means, now known or hereafter invented, including photocopying and recording, or in any information storage or retrieval system, without permission in writing from the publishers.

Trademark notice: Product or corporate names may be trademarks or registered trademarks, and are used only for identification and explanation without intent to infringe.

British Library Cataloguing-in-Publication Data
A catalogue record for this book is available from the British Library

Library of Congress Cataloging-in-Publication Data
Names: Rocheleau, Jordy, author.
Title: New interventionist just war theory : a critique / Jordy Rocheleau.
Description: Abingdon, Oxon ; New York : Routledge, [2022] |
Series: Routledge studies in intervention and statebuilding |
Includes bibliographical references and index.
Identifiers: LCCN 2021027411 (print) | LCCN 2021027412 (ebook) |
ISBN 9780367615284 (hardback) | ISBN 9780367615680 (paperback) |
ISBN 9781003105381 (ebook)
Subjects: LCSH: Just war doctrine. | Intervention (International law)
Classification: LCC U22 .R667 2022 (print) | LCC U22 (ebook) | DDC
172/.42–dc23
LC record available at https://lccn.loc.gov/2021027411
LC ebook record available at https://lccn.loc.gov/2021027412

ISBN: 978-0-367-61528-4 (hbk)
ISBN: 978-0-367-61568-0 (pbk)
ISBN: 978-1-003-10538-1 (ebk)

DOI: 10.4324/9781003105381

Typeset in Times New Roman
by MPS Limited, Dehradun

Contents

PART III
Just war procedures and application 167

Acknowledgements

I am indebted to many for their assistance in my completing this work. Austin Peay State University awarded me a Faculty Research Fellowship (aka Sabbatical) and Summer Fellowship, giving me time to plan the project. My colleagues in the Department of History and Philosophy have engaged in instructive conversations about military history and been supportive friends. Thank you in particular to Greg Zieren, Greg Hammond, Cameron Sutt, Dzavid Dzanic, and Bert Randall. Dewey Browder helped redirect my research to just war theory in 2001 when he invited me to a just war course in Austin Peay's new Military History Program. Thank you also to the many History and Philosophy students who engaged in thoughtful and challenging discussions in that course over the years.

Several friends have read chapters of this book and contributed valuable comments, including John Lango, Richard Peterson, David Chan, as well as three anonymous reviewers at Routledge. My understanding of issues in contemporary just war theory was advanced by a 2004 National Endowment for the Humanities Institute on War and Morality in the 21st Century in Annapolis. Discussion with colleagues at that seminar, including Bob Hoag, Harry van der Linden, Ed Santurri, Steven Lee, Michael Brough, Eric Patterson, Ken Rodman, John Lango, Gary Simpson, Mark Woods, and leader George Lucas was invaluable. My introduction to just war theory came as a graduate assistant in Steve Esquith's stimulating interdisciplinary course at Michigan State University; I am continuingly appreciative for this expansion of my applied ethics horizons.

My parents, Bruce and Georgette Rocheleau, offered ongoing encouragement throughout this project, as always. My deepest thanks go to my wife, Miyo Kachi, for her loving support for the duration of this long project, as well as acute editing assistance. Miyo and our son, Kaden, have been there for me throughout.

A version of Chapter 8 was previously published in the *Journal of Military Ethics*, Volume 19 (2), as "Legitimate Authority as *Jus Ad Bellum* Condition: Defense of a Procedural Requirement in Just War Theory." I appreciate the permission to republish it here.

Abbreviations

HRP	human rights paradigm
ICC	International Criminal Court
ICISS	International Commission on Intervention and State Sovereignty
ICJ	International Court of Justice
PAW	presumption against war
R2P	Responsibility to Protect
UN	United Nations
UNSC	United Nations Security Council
US	United States

Introduction

0.1 Just war theory and the new interventionism

Whether and when to resort to war is among the gravest and most pressing decisions facing communities and their leaders. Decisions about military force involve not only the prudential question of how best to pursue the national interest but also a moral question about the justifiability of killing and coercing in order to achieve one's objectives. Answers to the former prudential question are guided by conceptions of political and military strategy. The latter moral question has been systematically addressed by the tradition known as just war theory. From Cicero and Augustine to the present, theological and secular ethicists have defended general views about *jus ad bellum*, the moral justifiability of recourse to force, and *jus in bello*, the moral justifiability of the use of means in the course of war. There is considerable agreement among just war theorists on just recourse principles such as that uses of force must have a just cause, be a last resort, and be proportionate, as well as the central *in bello* principle of not targeting non-combatants. These moral criteria have been incorporated into the international law of war and the justifications for armed force given by political leaders and military decision-makers.

Nonetheless, the just war criteria are not settled doctrine. Just war theory faces ongoing challenges from both pacifists, who argue that war cannot be morally justified, and realists, who argue that war can and should be justified in prudential terms without ethical conditions. At the same time, there is debate within just war theory about the specification of its concepts and criteria, including what counts as a just cause, a last resort, and proportionate use of force. At least since the early modern period, just war theorists have debated whether war is justified to punish past crimes, prevent not-yet committed injustices, and stop moral abuses in foreign states. Just war theorizing occurs against an evolving political background, with shifting power, interests, threats, and opportunities. Social and political developments motivate new just war theorizing and provide a context to be taken into account in determining principles' desirability and practicability. Today, new threats and opportunities are motivating just theorists to

DOI: 10.4324/9781003105381-101

propose increasingly permissive criteria for armed humanitarian interventions, punitive attacks on groups and states associated with terrorism, and preventive strikes to quash gathering threats. Against this movement, this work defends a restrained view about the moral justifiability and practicality of using force for these putatively good aims.

In the aftermath of the World Wars, the twentieth century saw the crystallization of an ethical rejection of recourse to military force as ordinary political instrument. The UN Charter formalized a system of equal sovereign states among which armed intervention, except for defense, was condemned as aggression. Just war theory came to support this restrictive, "legalist" understanding of *jus ad bellum*. Paradigmatically, Michael Walzer's modern classic, *Just and Unjust Wars* (1977), defended a modified legalism in which only national defense and stopping the worst atrocities abroad are just cause for war. This view came to be widely accepted as just war orthodoxy. However, starting in the 1990s and continuing into the twenty-first century, the legalist framework has come under steady assault by new interventionist strains of thought. The trend in recent just war theory is to defend a reduced threshold for humanitarian intervention, wars to prevent gathering threats as well as preempt imminent ones, and the ready justification of both ongoing state-building occupations and small, targeted strikes, all in contravention to legalist orthodoxy.

There are various causes and rationales for this shift. Historically, the end of the Cold War and the dissolution of the Communist Bloc, and the stability they proffered, led to weakened states and the proliferation of civil wars. At the same time, there was a new opportunity for the US – with the help of Western Allies – to use its unrivaled superpower status to police the globe. This led to the rise of humanitarian intervention in the 1990s with support from sympathetic just war theorists. Increasing cultural and economic globalism has also contributed to military interventionism. Access to information about, including video clips of, the suffering of people around the world creates demand for humanitarian action. There is also a heightened sense of economic and political interdependence, with conflicts and human rights abuses in places such as the Balkans and Middle East creating refugee crises and market disruptions affecting the global community. The rise of terrorist groups with a global reach shockingly brought home to the US in 2001 created a greater sense of dangers from unstable states and non-state actors, leading to an interest in preventive intervention. In these circumstances, reasonable concerns about security and human rights and positive intentions for ameliorating these conditions fuel the new interventionism.

Even before the 1990s, the UN Charter's apparent ban on the first use of force for preventive and humanitarian purposes was perceived by ethicists as implausibly strict and morally indefensible (Luban 1980). Walzer's modified legalism, permitting intervention in exceptional cases, was reasonably charged with being insufficient in its communitarian ethical foundation and incoherent in its dual commitments to sovereignty and human rights, as well as still overly restrictive. A target of cosmopolitan-minded liberal critics from the beginning, the Walzerian orthodoxy has been largely rejected and displaced in the proliferation of just war thought since the end of the Cold War and the 9/11 attacks (Lee 2012).

Within just war thought, several theoretical approaches and numerous arguments have been used to justify lowered thresholds for armed intervention. Explaining and refuting these just war developments is the focus of this work. A first form of revision, led by hawkish neo-traditionalist scholar James Turner Johnson (1996), would return to classical understandings in which state leaders can readily employ military force to punish injustice and advance their conceptions of the good. The neo-traditionalists oppose the idea, implicit in Walzer and popular at least since its explicit formulation by James Childress (1992), that there is a presumption against war that must be overridden to ethically justify recourse to force. They also doubt that the last resort and proportionality criteria are strictly necessary and defend punishment as a just cause for war. Unbound by legalist strictures, the neo-traditionalists support most uses of force authorized by leaders of Christian, democratic states, including the US invasion of Iraq and its broader "War on Terrorism."

A second and increasingly dominant stream of just war thought takes a cosmopolitan approach, justifying war by the global responsibility to uphold human rights. Cosmopolitan hawkishness is paradigmatically displayed in the work of Fernando Teson (2005). It is also evident in the more measured theories of thinkers such as Allen Buchanan (2010), Steven Lee (2012), Cecile Fabre (2012), Brian Orend (2013), and John Lango (2014), among others. For these thinkers, legalism inappropriately fetishizes state sovereignty, the validity of which is conditional on human rights protection. Cosmopolitans conclude that human rights violations are a just cause for war. Unlike the neo-traditionalists, cosmopolitans condition war on universal principles. However, their rejection of the legalist framework has similar practical implications, potentially rationalizing preventive and policing interventions around the world. The cosmopolitans join the neo-traditionalists in rejecting Charter Law's requirement of international authorization and permitting the unilateral use of force to pursue moral goals. Some cosmopolitan thinkers have justified the invasion of Iraq (Teson 2005), while others argue that that adventure was a misapplication of their human rights paradigm (Bellamy 2014, 183).

In addition to these proposed general *jus ad bellum* (just recourse to war) frameworks, other arguments would also erode barriers to armed intervention. Many would reconceive just cause at a lower threshold. Some would replace the legalistic defense paradigm with a quasi-utilitarian norm in which war is justified if expected to further justice on balance. A variation of this argument asserts that small-scale interventions or wars otherwise planned to have a desirable benefit to cost ratio are justifiable at a reduced just cause threshold, blurring the line between war and policing. This argument would readily permit interventions such as those in Kosovo and Libya and counter-terrorist measures in assorted states throughout the world (such as Pakistan, Yemen, and Somalia) regardless of whether the injustices combatted meet the legal just cause threshold. Others argue that recent interventions have failed to further human rights and security because they have been too small and argue for a more thoroughgoing commitment to large-scale state-building (Orend 2013).

Another recent argument holds that multiple causes, each independently insufficient to justify war, can provide a just cause in combination. The ongoing intervention in Afghanistan and the invasion of Iraq following the 2001 terror attack on the US have both involved a mixture of defensive, humanitarian, and other goals, which are frequently presented as providing just cause in conjunction. Another war-excusing trend holds that once an intervention is underway, its continuation is presumptively justified until rights-preserving stability is achieved. Such a rationale lends support to continuing state-building operations in Iraq and Afghanistan, whatever one thought of the initial interventions.

Finally, there has been significant just war criticism of the principle of legitimate authority as a criterion for war. Interventionists argue either that legitimate authorization is unnecessary to justify war (Lee 2012; Lango 2014) or that, if it is required, it only entails national authorization (Brown 2011). This is in contravention to international law, which requires UN Security Council (UNSC) approval of armed interventions, and against the spirit of internationalism in tandem with which modern just war thought grew.

0.2 *Jus ad bellum* and the scope of this work

This work systematically analyzes and critiques these moves easing the justification of military intervention. Each leads to a dangerous proliferation of war and rests upon dubious inferences from and interpretation of just war principles. Against arguments that would lower the just cause threshold, I defend a modified legalism in which only national defense and humanitarian intervention to stop mass atrocities can justify war. I offer principled and practical arguments to bolster the maintenance of or, if we have already abandoned it, return to a version of legalist orthodoxy. On similar grounds, I argue against a reduced threshold for small wars, compound causes, and the default assumption of a war's continual justification. In response to cosmopolitan and neo-traditionalist critiques of internationalism, I defend a requirement of international authorization procedures, following international law and principled and pragmatic considerations of legitimacy in the use of force. Like other just war theorists, I reject a realism that justifies and limits war solely by concerns of national interest and power politics. However, I take seriously realist insights into practical limitations in our ability to use force for the good.

Permissiveness of armed force in the name of global security and human rights has, of course, been criticized by others. Pacifist writers have noted the potential for just war theory to become coopted into rationalizing and facilitating violence rather than restraining it (Yoder 1996; Fiala 2008). Ultimately, I find pacifism unpersuasive; its rejection of the possibility of just wars is not well-founded in principle and not as practical as a strict just war theory. My critique, then, is not directed at the idea of an ethics of war but rather toward its misapplication to justify war too easily. My position has some affinities with the more recently developed contingent (or "conditional" or "practical") pacifism, which argues that although war in principle could be justified, contemporary wars cannot be or are very unlikely to be justified, such that there is reason to oppose them in general (May 2015). I share the view that taking just war principles seriously means viewing war more skeptically and with greater restrictions than many of today's theorists and practitioners. However, like other traditional just war theorists, I think that war sometimes can be justified for national defense or to prevent mass atrocities abroad, despite armed conflict's inherent costs and risks. This work develops an immanent criticism of just war theory's recent trajectory and seeks to restore Walzerian legalism's balance of intervention to stop the worst injustices with the prohibition of its use in ordinary circumstances.[1]

This work is about *jus ad bellum*, the justice of recourse to war. I leave aside other issues in just war theory and military ethics, especially *jus in bello*, justice in the means of fighting, except insofar as the means affect considerations about whether to use force at all.[2] I also am not centrally concerned with the recently theorized *jus post bellum*, insofar as it addresses the justice of punitive and reconstructive measures after wars. However, end-of-war arrangements are relevant insofar as they affect the viability of war as a method of promoting justice. I also address the question of how long parties justly can and should continue wars and post-war occupations, an issue sometimes discussed under the *post bellum* rubric. My argument in Chapter Seven that wars must continually satisfy *jus ad bellum* is relevant to assessing the justification of the long interventions in Afghanistan and Iraq and other ongoing peacekeeping missions.

Having previewed the major issues to be covered and theses to be defended, let me introduce the just war framework and method I employ to discuss the justification of recourse to war.

0.3 *Jus ad bellum*: initial principles

I begin reflections on the just recourse to war with principles long held in the just war tradition to govern the resort to force. Principles including those of just cause, right intention, legitimate authority, last resort, and proportionality are supported by intuition, consideration of consequences, and other more general principles and duties. They have also been incorporated into international law, further evidence of their plausibility and practical applicability. Most disputes among just war theorists and international lawyers are not about these principles' validity but their interpretation and practical application.

To decide among interpretations of principles such as that of just cause, it is necessary to refer to other principles and arguments which underpin our conflicting judgments. One sort of consideration is deontological, that is, in terms of the logic of duty or obligation. Ross (2012) argued that there are intuitively known *prima facie* (in principle, other things being equal) duties, which include not harming others, being just (i.e. fair), being truthful, and redressing harms for which one is responsible. Such general principles underpin most of our moral judgments.

For resolving dilemmas, when the principles conflict, we have to go beyond the intuitionism to which Ross appealed. In such cases, we can employ a universalizability test, which Kant (2006) called the "categorical imperative," wherein moral norms are only valid if one could have all follow them. An action or policy chosen out of self-interested exceptionalism is unjust. For example, a preventive war might be desirable for an individual party's security, but security would be undermined if all adopted the policy of attacking preventively. That it would be self-defeating if others committed similar acts and that one would not want them to follow the policy condemns it as unjust. Admittedly, there are many objections to Kantian ethics. Consistent universalizability is not sufficient to establish moral desirability. In principle, a party might consistently will that all engage in coldly rigid non-assistance or bellicose war-mongering so long as it did not desire assistance or security itself. Insofar as universalization is based on an agent's intentions, it has an element of subjectivity and requires interpretation, making its implications unclear. Nonetheless, it presents a strong case against a view of the ethics of war if one would not generalize the recommended principle.[3]

Many arguments about the morality of war are couched in terms of "human rights," minimum claims that people have against one another, which are valid independently of individual interests or even the greater good. The idea of a duty to other persons such as ourselves arguably underpins much of just war theory, including both the *in bello* rule of discrimination (or non-combatant immunity) and *jus ad bellum* principles such as just cause and last resort. Various foundations of human rights have been offered. From the categorical imperative discussed above, Kant derived the corollary that we have a duty to respect human beings, treating them as "ends in themselves and never merely as means." Some contemporary thinkers derive human rights from the conditions required for rational agency or a decent human life (Griffin 2010). Others find them to be grounded in international recognition and political practice (Raz 2010). For our practical purposes, we do not need to resolve these debates about foundations, as in practice, they imply similar respect for human life and basic liberties.

Human rights are not without controversy. Some doubt their universality, contending that they are conceptually tied to presuppositions of the western individualist and Judeo-Christian traditions (Runzo, Martin, and Sharma 2003). Even if one accepts them in principle, difficulties about just which norms count as human rights arise. There is controversy about how to draw the line between rights and other things which are desirable but whose denial is not a human rights violation. For example, many have questioned whether every claim acknowledged in the Universal Declaration, such as a right to "periodic holidays with pay" (United Nations General Assembly, 1948), absolutely must be recognized and protected by all societies.

Nonetheless, the idea that there are minimal standards of decent treatment that should be accorded to all people is eminently plausible and has become the closest thing to a fundamental principle in just war thought. Even Walzer, who is generally viewed as being on the collectivist and non-foundationalist end of the just war spectrum, acknowledges a priority of rights claims (1977, 54). As discussed above, the cosmopolitan view that the protection of human rights is necessary and sufficient to justify war has become predominant in recent just war thought.

I agree that there is a strong presumption against violating human rights claims and view the greater protection of human rights as a central goal in international relations. However, I would also argue that human rights concerns have been used overly abstractly and dogmatically to justify armed intervention without regard for context, consequences, and competing obligations. While the prohibition of violating human rights is near absolute, the use of military force to prevent any and all rights violations is problematic. War itself inevitably infringes rights, including especially the most basic right to life, and undermines other values such as peace and security on a large scale. On a plurality of grounds, I will argue that it is a mistake to accept human rights violations short of mass atrocities as a just cause for war. In decisions where both recourse to and refusal to employ force will result in rights violations, the best we can do is employ a sort of "utilitarianism of rights." This means attempting to protect rights as much as possible, taking into account the relative weight of different rights and the deontological priority of not violating rights oneself over stopping rights violations by others. I will argue that human rights are, only seemingly paradoxically, best promoted by not permitting wars to promote them except in the most egregious cases.

0.4 Consequences, rules, and the morality of war

I have argued that there are general moral duties that are relevant to the waging of war. However, we also saw that there is disagreement about these principles' exact nature and implications. It is doubtful that any are absolute, as the duties may conflict, or following them strictly might be excessively costly and impractical. For all of these reasons, moral judgment needs to take into account factors other than deontological principles.

Among the factors that must be considered are the practical effects of actions and practices. In ethical theory, the main competitor to deontology is consequentialism, in which the desirability of its results determines an action's morality. Paradigmatically and most influentially, utilitarianism holds that one should act so as to maximize happiness. Producing positive consequences appears to undeniably speak in favor of an action, and producing worse consequences than would otherwise pertain tells strongly against it. This straightforwardly applies to decisions to go to war (as well as actions in the course of war). The generally accepted *jus ad bellum* proportionality principle, which rejects wars that would do more harm than good, directly imposes a utilitarian condition on the recourse to war.

However, consequentialism is implausible as a comprehensive moral theory and as a basis for the ethics of war in particular. An action could maximize the good but still be wrong, as in the classic example of killing an innocent person to harvest his or her organs to save several dying individuals. In just war matters, an intervention to overthrow a foreign government could be beneficial to humanity on balance, say by promoting local and global economic development, but would still be unjustified if it lacked just cause or was not a last resort. Recourses to violence that violate basic moral principles are generally immoral regardless of the expected consequences. This is why cost-benefits proportionality is necessary but not sufficient to justify war.

A more plausible consequentialism, "rule utilitarianism," holds that instead of always performing the action that will lead to the greatest good, one should act according to those rules which will produce the most good if generally followed. This combines the insight about the importance of consequences with a universalization test in which a party must consider the desirability of all doing as it does. The rule utilitarian test further explains the justification of the *jus ad bellum* principles and aids in their rational interpretation. It tells strongly against a conception of just cause or legitimate authority if its implementation would have relatively harmful effects. I will argue that this is the case with new interventionist concepts of just cause, including cosmopolitan human rights enforcement, preventive war, and compound just causes. Although I do not defend rule utilitarianism as an absolute moral foundation, I find that it provides a reasonable adjudication between principle and consequences in many cases.

Rule utilitarian concerns are related to another normative source: positive law. Norms that are legally enacted and recognized generally ought to be observed. Regarding *jus ad bellum*, the international law of the UN Charter prohibits the waging of non-defensive wars, at least without UNSC authorization. National laws put further procedural strictures on the recourse to force. Of course, legality does not subsume morality; an action could be legal but unjust or illegal but justifiable. The law at any time may be mistaken or, if generally just, may have ethical exceptions. Moreover, to the extent that the law is just, it is presumably based on other ethical rationales, such as the principles and consequences discussed above. Thus, legality could be argued to be less basic than those other moral considerations. For all these reasons, many just war thinkers, especially the interventionists criticized here, discuss the morality of war independently of legal considerations.

Against this trend, I argue that a theory of just recourse must take into account the law of war. As widely promulgated, generally accepted, and to varying extents enforced norms regarding warfare, the international law of armed conflict is a reference point for any just war theory which seeks to be relevant to actual conduct. War law has been shaped by ethical arguments from just war theory, such that it is a rationally tested candidate for what is morally acceptable in the use of force.[4] Moreover, insofar as parties are committed by contract to legal norms, there is a *prima facie* obligation to obey the law. Finally, I argued above the consequences of types of actions should be taken into account. Just war decisions are made in a world in which recognized legal norms function to maintain international security and stability. Violations of norms prohibiting the use of force, which might be beneficial in an individual instance, can be evaluated by rule utilitarian calculations regarding what constitutes a valid action guiding norm. Unlike domestic uses of force by individuals, international uses of force and the international community's response to acts of war help to establish (or undermine) customary international law. For all of these reasons, legality is a factor in the ethical evaluation of war, albeit not the only factor. The laws of the UN Charter and the recently adopted R2P norm explicitly permitting authorized armed interventions to stop atrocity crimes (ICSS 2001), become a good starting point for an ethics of war, which I argue is in turn supported by both consequentialist and deontological considerations. It is against this modified legalist framework that new interventionist proposals should be tested.

Another influential and useful way to think about the ethics of warfare between groups is through a "domestic analogy" to norms governing violence between individuals in civil society (Walzer 1977, 58–59). Although not an international legal norm, the domestic analogy is an important source of the morality of states and other groups. Because our moral and legal language is most developed regarding the relations between individuals, it is almost inevitable that we think of group conflict in similar normative terms. In many respects, states are treated as individuals, as we refer to them as having intentions, willing, making treaties, being legally bound, and having rights to self-determination and physical integrity, as agents akin to persons. Thus, there is *prima facie* reason to apply norms regarding individuals in domestic society to states in international society.

Nonetheless, there are disanalogies between states and persons. States are abstract entities and do not possess the singular consciousness that gives individual humans moral status. Moreover, states are composed of many persons. On the one hand, this means that states acquire moral value from their furtherance of, if not necessity for, numerous individual needs, interests, and values. On the other hand, states can (and frequently do) violate their own citizens' rights in the course of actions that are within the scope of state self-determination from an international standpoint. Thus additional moral analysis is required to adjudicate between national sovereignty and human rights. The less reliable enforcement of international as opposed to domestic law can also be argued to reduce the validity of the former's norms and parties' obligation to comply. Such practical and contextual considerations must be addressed when making inferences about international norms from individual morality.

If the above is correct, *jus ad bellum* must take into account the consequences of military actions and their relationship to norms and laws governing war, along with the deontological principles discussed in Section 0.3. The question becomes how to combine and adjudicate between these different considerations.

0.5   Ethical pluralism and a reasonable just war theory

Ethicist George Lucas (2013) has decried a crisis of "methodological anarchy" in just war theory in which theorists take conflicting, isolated approaches, some legalist, some theological, others based on the classic just war tradition, and others (like Walzer) based on case analysis. Although this phenomenon is real, I would suggest that the problem is not that there is too wide a number of considerations but that each is too often portrayed as if it is a foundation from which to mechanically derive moral conclusions without dialoguing with the others. Indeed, it is my view that mistakenly abstract moral foundations have led to overly permissive proposals regarding the international use of force. Above, I argued that multiple considerations are relevant to the justification of war, with none being absolute, as any are potentially overridden by the others.

The question becomes how to integrate the considerations for determining just recourse to war. Ideally, there would be one fundamental principle or at least a systematic overarching rule from which right conduct can be derived. However, I have given reasons to think that this is not the case. This points to an ethical pluralism in which various principles are all taken into account.[5] The challenge for a pluralist view is to show that the integration of the principles will be rational, based on reasoned argument, rather than merely arbitrary. I outline such a non-foundationalist approach here.

Like most just war theorists, I begin with widely accepted *jus ad bellum* principles. Since the just cause criterion is rarely challenged, I focus on disputes about its interpretation. I adjudicate between various interpretations by testing them against the deontological and consequentialist considerations discussed above. In the case of legitimate authority, as the criterion itself has been frequently questioned, I employ principled and practical arguments to defend its validity against objections. To the extent that all of the considerations support a principle or interpretation, the latter becomes strong. When it is supported by some ethical arguments but also has important objections, there is an ethical dilemma, and an adjudicating judgment must be made. I will sometimes deploy one of two meta-principles introduced above to rationally adjudicating conflicts between deontological principle and consequentialist pragmatism: either rule utilitarianism or a utilitarianism of rights. Nonetheless, not even these principles provide an unquestionable foundation, such that contravening moral considerations must be taken into account.

Making judgments about disputed cases inevitably draws upon our intuitions about the relative weight of moral arguments and the plausibility of the implications of principles for particular cases. When moral principles conflict with our intuitions about the right course of action, we can either revise our principles or revise our judgment about the case. For example, if we think that preventive war is wrong in principle, but we favor it in a particular instance, reason demands the reconciliation of this conflict. This implies a sort of justificatory circle in which principles are supported by intuitions, but the intuitions are supposed to be based on the principles. Such a circle is inevitable in ethics, where preliminary normative assumptions need to be made to begin reflection and argumentation. However, the circle can avoid arbitrary dogmatism if we treat neither principles nor intuitions as fixed. The process of working back and forth between moral intuitions and moral principles until they are mutually supporting has been well described by John Rawls (1971) as one of attaining "reflective equilibrium." Because it involves testing for consistency and coherence and takes into account the relative weights of different considerations, such a non-foundational pluralism is nonetheless rational.

Insofar as Rawlsian reflective equilibrium suggests a process that occurs within the mind of an individual thinker, I would combine it with the on-going debate in which one must answer objections from opposing perspectives. Ultimately it is the human community that determines what is morally justified. A view of the justice of intervention and preventive and punitive wars will be derived from an assessment of arguments regarding the meaning of just cause, the implications of human rights, and the consequences of various action plans. A proposal must be tested and defended against alternative views and objections regarding the implications and weight of these and other considerations.

Ideally, a consensus on action-guiding principles can be reached by persuading parties from various perspectives. If all parties do not agree for the same reasons, it may nonetheless be possible to find what Rawls (1993, 39) called an "overlapping consensus," in which a principle of conduct is supported for different reasons by parties with different values. For example, the condemnation of aggression might be simultaneously supported by

Kantian, utilitarian, theological, and legal arguments. While actual moral debates will probably never be finally resolved, with conclusions unquestionably fixed, proposals that survive rational testing thereby acquire a degree of confirmation.[6] A proposal that responds to and is acceptable from various standpoints is considered "reasonable" in its defensibility, as well as internally rational. Thus a non-foundationalist, pluralist approach to the justice of war should ultimately lead to conclusions that can rationally and reasonably guide conduct.

0.6 Plan for the book

This book systematically analyzes and critiques developments in just war theory which lower the bar for justifying recourse to war. Although these interventionist just war views are reasoned, I argue that they are ultimately implausible. They use abstract principles to derive an ethics of war that is dangerously permissive. I will argue that a pluralist approach, taking into account international law and the practical context of modern warfare as well as principled conditions for the justice of force, points toward a more stringent, albeit modified, legalist paradigm with a high threshold for intervention and a robust requirement of international authorization.

The first part of the book defends the necessity of meeting just recourse criteria to override the presumption against war. In Chapter One, I defend the just war approach against realism and give an initial exposition, defense, and interpretation of the *jus ad bellum* principles. In Chapter Two, I respond to the neo-traditionalist proposals that there is not a presumption against war and that the satisfaction of all the *ad bellum* criteria is not necessary. This view derives from a combination of a hawkish interpretation of the just war tradition with a permissive view about the prerogatives of sovereignty. I reject this interpretation, the inferences drawn, and the view of sovereignty and argue that there is a presumption against war and war that can only be overridden when all the *jus ad bellum* principles are satisfied.

In Chapter Three, I take up the neo-traditionalist resurrection of the Augustinian idea that punishment is a just cause for war. Legalism and enlightenment morality had undermined the view that war can be undertaken for the purpose of punishing evildoers. However, disgruntlement with the restrictiveness of legalism combines with concern to respond to global terrorism and states condemned as rogues. Against this move, I attempt to demonstrate that neither moral principle nor consequentialist considerations, nor even a combination of both, can justify punishment as an independent cause for war.

The second part of the book discusses the just cause threshold. Chapters Four and Five respond to the most influential challenge to the legalist paradigm, the cosmopolitan view that human rights protection is just cause for war. In Chapter Four, I lay out the cosmopolitan argument that the human rights paradigm is morally better founded and internally more consistent than the national defense paradigm. Against this emerging consensus, I argue that the human rights paradigm is itself riddled with ambiguities (such as those between rights violations, loss of legitimacy, and just cause for war), has an implausible view of legitimacy and sovereignty, and does not offer a sufficient justification for war. I also take up and reject the argument that foreign intervention is justifiable whenever revolution is. This chapter gives an initial defense of the need for a legalism with a high just cause threshold.

In Chapter Five, I take up the consequences of intervention. While cosmopolitans and other interventionists generally accept the proportionality criterion, their permissively low just cause threshold invites wars that are apt to have poor consequences, including for human rights. I describe unintended consequences that tend to make wars more harmful than expected and epistemic challenges for parties evaluating whether a war to which they are inclined satisfies just war principles. The chapter critically responds to arguments that by committing more robustly to nation-building on the one hand or cautiously engaging in small, targeted interventions on the other, war can be used more readily to advance human rights. While cosmopolitans tend to see war as continuous with ordinary law enforcement, I defend the significance of the war/policing boundary, with the *ad bellum* criteria limiting the former and strict rules of engagement limiting the latter. In addition to problems likely in any particular intervention, replacing the legalist non-intervention norm with one of interventionist humanitarianism would cause systemic problems. Rule utilitarian and international legal and legitimacy considerations further the case for a defense paradigm with exceptions only for atrocities.

Chapter Six takes up a new pragmatic proposal for justifying wars that lack any single clear just cause: the idea of a compound just cause. Frequently, there is more than one rationale for any war under consideration, such as the varied justifications given for the ongoing US intervention in Afghanistan and the (notoriously) numerous causes offered for invading Iraq. Such long lists of rationales suggest that none of them is in itself sufficient to justify war. However, in such cases, interventionists sometimes argue that the independently insufficient causes can form a just cause in conjunction. While I acknowledge that there are ways in which multiple purposes can add to the justification of war, I argue that compound causes are generally invalid and that war requires a unitary, independently justifying cause. My argument helps to clarify why the Iraq War lacked just cause and was thus unjustifiable and casts doubt upon continuing operations in Afghanistan.

Part Three takes up procedural issues in the justification of war. Chapter Seven addresses the question of how to evaluate ongoing interventions over time. It responds to arguments from Orend and Thomas Hurka that would allow the continuation of occupation and peacekeeping without necessarily having a just cause for war or intervention still being last resort. A recently introduced concept of *jus ex bello* continues this tendency to treat the problem as one of whether to end a war rather than whether to continue it. I argue that wars must meet the criteria for just recourse throughout, including during occupation after the initial assault is over. The prolonged interventions in Afghanistan and Iraq and the recent debates about state-building make pressing this chapter's theory of when it is permissible to continue and obligatory to end wars.

In Chapter Eight, I explain and defend the principle of legitimate authority. Recent proposals would either do away with the right authority criterion or limit the requirement to national authorization by the belligerent's own government. I counter by defending the moral necessity of proper authorization and argue for, in particular, the primacy of international authority with exceptions only through rational, universally justifiable procedures. I note the implications of this for war by non-state groups, dubiously legitimate states, and parties who would conscientiously disobey international law.

The conclusion, Chapter Nine, takes up the question of my anti-interventionism's practical implications for the threats to human security around the world today. Would intervention be justified in Syria and other states with serious ongoing human rights abuses? Are preventive strikes against rogue states and non-state actors which pose threats justified? What of targeted strikes short of full-scale war? In response, I reject most interventions and all preventive wars. However, I note that my view permits intervention in some of these cases, contingent on specific empirical factors. I give a qualified defense of some targeted strikes as actions short of war; however, I recommend more strict rules even here than have been applied heretofore. I conclude by arguing that an internationalist, non-interventionist just war theory such as mine could effectively guide practice.

Notes

1 Similar concerns about the easy justification of war for human rights have been raised by other just war theorists. I draw on critiques of interventionism by theorists such as A.J. Coady (2012) and Richard Miller (2003).
2 There are certainly important issues in the implementation and interpretation of *jus in bello* as well, including defining combatancy and individual liability to attack, the status of POWs in the "war on terror," and the meaning of proportionality. The spread of enhanced interrogation, which amounts to torture, the use of the doctrine of double effect to readily permit killing non-combatants, and urges to expand the scope of who counts as a combatant all call for attention to *jus in bello* in order to properly restrain the use of force.
3 Some defenders of intervention and preventive war reject the application of a norm of universalization (Buchanan 2007). I reply to this view in Chapter Five.
4 Larry May (2008) analyses the *ad bellum* principles underlying the crime of aggression.
5 My non-foundationalist pluralism, in which multiple ethical factors are taken into account, contrasts with an alternate conception of ethical pluralism in which various

moral frameworks independently justify actions without taking other moral values and principles into consideration. The latter communitarian view is rooted in respect for diverse moral perspectives but is insufficiently rational in forswearing the possibility of reasonable adjudication between values and principles.

6 Proponents of communicative or discourse ethics, paradigmatically Habermas (1990), argue that moral truth is constituted by rationally motivated agreement. For our practical purposes, it is not necessary to resolve whether rational debate creates morality or whether it uncovers pre-existing moral truths. Either way, rational dialogue is presumptively the approach to finding justifiable moral principles.

References

Bellamy, Alex. 2014. "The Responsibility to Protect and the Problem of Regime Change." In *The Ethics of Armed Humanitarian Intervention*, edited by Don Scheid, 166–86. Cambridge, UK: Cambridge University Press.

Brown, Davis. 2011. "Judging the Judges: Evaluating Challenges to Proper Authority in Just War Theory." *Journal of Military Ethics* 10 (3): 133–46.

Buchanan, Allen. 2007. "Justifying Preventive War." In *Preemption: Military Action and Moral Justification*, edited by Henry Shue and David Rodin, 126–42. Oxford: Oxford University Press.

Buchanan, Allen. 2010. *Human Rights, Legitimacy, and the Use of Force*. Oxford: Oxford University Press.

Childress, James. 1992. *Moral Responsibility in Conflicts*. Baton Rouge: Louisiana State University Press.

Coady, C.A.J. 2012. *Morality and Political Violence*. Cambridge: Cambridge University Press.

Fabre, Cecile. 2012. *Cosmopolitan War*. Oxford: Oxford University Press.

Fiala, Andrew. 2008. *The Just War Myth*. Lanham, MD: Rowman & Littlefield Publishers.

Griffin, James. 2010. "Human Rights and the Autonomy of International Law." In *The Philosophy of International Law*, edited by Samantha Besson and John Tassioulas, 339–56. Oxford: Oxford University Press.

Habermas, Jurgen. 1990. *Moral Consciousness and Communicative Action*. Cambridge, MA: MIT Press.

ICISS. 2001. "The Responsibility to Protect." http://responsibilitytoprotect.org/ICISS%20Report.pdf

Johnson, James Turner. 1996. "The Broken Tradition." *The National Interest* 45: 27–37.

Kant, Immanuel. 2006. "Groundwork of the Metaphysics of Morals." Reprinted in *Ethics*, 3rd edition, edited by Steven Cahn and Peter Markie, 270–308. New York: Oxford University Press.

Lango, John. 2014. *The Ethics of Armed Conflict*. Edinburgh: Edinburgh University Press.

Lee, Steven. 2012. *Ethics and War*. Cambridge: Cambridge University Press.

Luban, David. 1980. "Just War and Human Rights." *Philosophy & Public Affairs* 9 (2): 160–81.

Lucas, George. 2013. "'Methodological Anarchy' – Arguing About Preventive War." In *Empowering Our Military Conscience: Transforming Just War Theory and*

Military Moral Education, edited by Roger Wertheimer, 33–55. Hampshire, UK: Ashgate Publishing.

May, Larry. 2008. *Aggression and Crimes Against the Peace*. Cambridge: Cambridge University Press.

May, Larry. 2015. *Contingent Pacifism*. Cambridge: Cambridge University Press.

Miller, R. 2003. "Respectable Oppressors, Hypocritical Liberators: Morality, Intervention, and Reality." In *Ethics and Foreign Intervention*, edited by Deen Chatterjee and Don Scheid, 215–50. Cambridge: Cambridge University Press.

Orend, Brian. 2013. *The Morality of War*, 2nd edition. Buffalo, NY: Broadview Press.

Rawls, John. 1971. *A Theory of Justice*. Cambridge, MA: Harvard University Press.

Rawls, John. 1993. *Political Liberalism*. New York: Columbia University Press.

Raz, Joseph. 2010. "Human Rights without Foundations." In *The Philosophy of International Law*, edited by Samantha Besson and John Tassioulas, 321–38. Oxford: Oxford University Press.

Ross, William David. 2012. "The Right and the Good." In *Ethics*, 5th edition, edited by Steven Cahn and Peter Markie, 475–84. New York: Oxford University Press.

Runzo, Joseph, Nancy Martin, and Arvind Sharma, eds. 2003. *Human Rights and Responsibilities in the World Religions*. Oxford: Oneworld Press.

Teson, Fernando. 2005. "Ending Tyranny in Iraq." *Ethics & International Affairs* 19 (2): 1–20.

United Nations General Assembly. 1948. "Universal Declaration of Human Rights." Resolution 217A. https://www.un.org/en/universal-declaration-human-rights/

Walzer, Michael. 1977. *Just and Unjust Wars*. New York, NY: Basic Books.

Yoder, John Howard. 1996. *When War is Unjust: Being Honest in Just War Thinking*. Maryknoll, NY: Orbis Books.

Part I

Ethical constraints on recourse to war: *jus ad bellum* vs realism and neo-traditionalism

1 Just war theory and the ethical restraint of war

1.1 The triumph and limits of just war theory

In the modern world, governments cannot undertake military operations without defending them as morally justified. These justifications, in turn, inevitably use terms such as "just cause," "last resort," and "proportionality," derived from just war theory. As its terminology has become part of the public vernacular, the just war discipline has developed extensively in the years since Walzer's seminal *Just and Unjust Wars*. Analytical philosophers provide increasingly precise accounts of the rights and responsibilities at stake in warfare (McMahan 2009; Lango 2014).

However, this "triumph of just war theory" (Walzer 2004) has not succeeded in preventing dubious armed interventions and the proliferation of preventive and punitive strikes. In practice, a moral argument is limited in its ability to overcome realist and militarist impulses. Leaders face pressures to act quickly and decisively when considering military force, without the constraint of moral reflection. For the public, existential threats promote a narrowing of concern to self and community, crowding out critical analysis and a broader sense of moral obligation. During an international conflict, the publics tend to rally to support their own forces regardless of their prior independent moral judgments of the merits of the case. Politicians, themselves under nationalistic sway, are eager to respond to a public demanding strong action and condemning weakness. Bill Clinton learned as president that it is "better to be strong and wrong than weak and right," a lesson that seems to be internalized by each new administration and Congress.[1]

Insofar as *jus ad bellum* terminology is used to justify the use of force, it may amount to moral "window dressing" for states and leaders to rationalize their self-interested *realpolitik*. At the same time, as I outlined in the Introduction, just war theory has, in turn, shifted in directions that make it conducive to the justification of force. Critically responding to this just war interventionism is the central theme of this book. However, before we get to those debates, I note that intellectually, as well as in unstated political motives, realism remains an influential competitor for just war theory. Among military and political strategists and historians, it remains the main

DOI: 10.4324/9781003105381-1

lens of analysis. Realism has received a recent cultural boost, with the rise of nationalist movements reacting to discontents with cosmopolitan liberal universalism. Realists argue either (1) that, factually, states and other international actors are motivated by their own interests rather than moral considerations in international relations and the use of force in particular, or (2) that, normatively, states *should* act according to national interest rather than moral principles in international relations, including the use of force (Coates 2007). Many accept a combination of factual (or descriptive) and normative (or prescriptive) realism. For reasons I articulate in the next section, both realist theses should be rejected.

1.2 Realism and the justification of war

The starting point for just war theory is that war is only justified if it satisfies ethical conditions. Thus, just war theory is essentially opposed to realism's assertion that warfare must be guided by power and interest rather than morality. Because of realism's historic prevalence, its ongoing pull, and its relation to just war theory, it will be helpful to explain the realist view and why it must be rejected here.

Variations of realism have been defended by numerous thinkers, including political philosophers Machiavelli and Hobbes, historians E.H. Carr and John Mearsheimer, political scientists Hedley Bull and Hans Morgenthau, and diplomats Henry Kissenger and George Kenan. Its classic statement comes from the *History of the Peloponnesian War* in which Thucydides has Athenian generals justify their takeover of Melos and threatened (and ultimately executed) genocide against the Melians with the statement that in war, "the strong take what they can and the weak suffer what they must."

Realists frequently cite military strategist Carl von Clausewitz's dictum that "war is nothing but a continuation of politics by other means" to explain their view that a state is justified in going to war when and only when doing so furthers its own interests.[2] A preventive war to weaken a potential adversary or replace its government with one more friendly to one's own aims – which ordinary morality would condemn as aggression – would be justified on the realist view so long as the prospective benefits *to the aggressor* outweigh the costs.[3] For realists, the US invasion and occupation of Iraq was justified if it could have been expected to further US interests, independently of the culpability of the Hussein regime or the welfare and will of the Iraqi people.[4] On the other hand, a humanitarian intervention, like that in Kosovo or a proposed one to stop the genocide in Rwanda, should not be undertaken if it could not be expected to benefit the *intervener* on balance.

Several arguments are used to support realism. A frequent rhetorical move, with a kernel of underlying argument, is that war is a realm of "necessity" rather than choice. This implies that when it comes to life and death struggles, we must react automatically, without pausing for principled reflection and analysis. War is entered under duress so that moral choice and responsibility are impossible and inapplicable. Indeed, war frequently seems inevitable, especially as we look at it in hindsight as the product of an escalating conflict.

Nonetheless, it is simply false to describe wars as outside the realm of choice. Groups of people decide to go to war. Marshaling and deploying forces takes planning and deliberation. Moreover, the use of force is usually recognized as a weighty decision and undertaken after consideration of its likely effects and possible alternative courses of action. Strategic debate among realists about the timing and manner of fighting wars shows that even they view war as a matter of rational deliberation and not outside human agency. As Walzer (1977, 8) comments in his critique, realism frequently trades on the ambiguity of the term "necessity," which "doubles the parts of inevitability or indispensability." War, as I just argued, is never inevitable. It might be indispensable for certain ends, but this assumes there are no other feasible means to those ends. We can also ask whether the ends, such as preserving or capturing a piece of territory or weakening a rival, are themselves morally indispensable – or otherwise sufficient to justify war. Reflection on these matters puts us in the ethical realm of just war theory, so the rhetoric of "necessity" cannot consistently establish realism as an amoral alternative.

A second argument for realism derives from a social contract conception of the purpose and limits of government. Hobbesian realists argue that states are formed to further the rights and well-being of their citizens. If a government pursues the welfare of or even takes care to respect the rights of foreign nationals, it oversteps its mandate and contravenes its purpose.

However, the pursuit of national interest is not the sole purpose of government, to the exclusion of all considerations of the lives of non-citizens. Even if – as seems eminently plausible – furthering the welfare of the state itself is a *predominant* purpose of government, it cannot justify violating the rights of foreign citizens. Although realists write as if moral obligations have no place in war, the realist doctrine itself presupposes the moral responsibility to maintain the promises implicit in the social contract, which is to say it presupposes the moral validity of promises prior to the establishment of government. Yet, if people have the right to have contractual promises kept, they certainly also have other rights, including the rights to life and liberty. The principled realist rejection of obligations to foreign nationals is contradictory and unjustifiable.

Furthermore, as others have noted, the realist's central concept of a "national interest" to be pursued in international relations is rarely defined and collapses into ambiguity upon analysis. Realists sometimes write as if the national interest is limited to state survival, a ridiculously unambitious view of the aim of government. In fact, national survival is rarely at stake in international relations. For example, an intervention in Rwanda would have cost the US and Europeans some money and casualties but would not have placed their existence in jeopardy. States with relatively generous foreign aid policies, such as Canada, do not thereby place themselves in existential crisis. On the other hand, although the US military response to Pearl Harbor and the 9/11 terror attacks are thought of as paradigmatically justifiable acts of self-defense, the refusal to respond with force probably would not have led to the overthrow of the United States. A broader, somewhat more plausible view of the national interest adds other material interests, especially furthering the state and its citizens' economic interests. However, a purely materialistic view of national well-being is still overly narrow, as people have values and ideals beyond financial success. Citizens generally do not expect their governments to maximize their material welfare at the expense of all moral obligations, particularly those of human rights to life and liberty, nor would they be justified if they did. Any plausible view of the national interest includes the pursuit of moral values and principles, which in turn includes not waging war without ethical justification (Welch 2000).

A second Hobbesian argument for amoral realism asserts that, given the anarchical state of international relations, nations cannot safely act upon any principle other than self-interest. Thus, even if one thought it desirable to have states take into account each other's well-being, one might think that, without assurance that other states will reciprocate, it is too dangerous to act on these moral ideals. Without a world government enforcing inter-national norms of non-aggression or requiring mutual assistance, each state is susceptible to attack by others.

However, it is not correct to depict international society as a Hobbesian state of nature. States do not risk their own existence by failing to take advantage of others' weaknesses or by making small sacrifices of their own interest for the benefit of others. This is partly because of their geographic separation and relative independence – states are not as vulnerable to attack as individuals in an anarchical territory. More basically, the international system is not anarchical. There is a well-developed body of international law governing armed conflict (Cassese 2005). States recognize moral and legal obligations to each other. They can be taken to the ICJ and sanctioned for violations, and their leaders can be tried at the ICC. Although sanctions are less regular in international than domestic law, violations of treaty obliga-tions and other norms undermine a state's standing and can lead to less favorable international treatment, such that norms have teeth as well as moral force behind them. In response to aggression, states can exercise "collective defense" by joining on the side of an attacked party, as Allied forces did in response to Iraq's invasion of Kuwait and North Korea's in-vasion of the South. The UNSC can also authorize armed intervention in response to "threats to international peace and security" short of aggression.

Contrary to realism, it is more accurate to say that states cannot afford to ignore their moral obligations than that they cannot observe them. States regularly take into account morality in their international relations and military decision-making. The common practice of financial assistance to other states suggests this. Recent interventions, such as those in Somalia, in Serbia over Bosnia and Kosovo, and most recently in Libya, appear to be motivated in large part by the desire to provide humanitarian assistance. In general, as Walzer argues, political and military officials do not justify their actions in terms of power politics but rather provide moral justifications and attempt to defend their actions against moral criticisms. They are forced to explain why their wars are "necessary" or, in any case, "justifiable" terms that refer to moral principles. This is especially true of democratic states but holds even for authoritarian states, which have to maintain the ideological support of their people, as well as a degree of international cooperation, to function. Of course, these defenses are sometimes cynical rationalizations for selfishly or nationalistically motivated military acts. Yet, even moral window-dressing shows the necessity and efficacy of moral discourse, the requirement of attempting to persuade the public of one's moral war plans. Not everything can be defended as moral. The more people are able to engage in the analysis of arguments regarding the morality of war, and the more they demand their governments fight morally, the more efficacious moral arguments become. International behavior and concepts of moral justification make realism implausible (Walzer 1977, 15–20).

There is a more moderate, reasonable form of realism. This theory is skeptical about our ability to achieve moral norms in war and argues that the pursuit of seemingly moral or just goals tends to cause more harm than good. This sort of realism is found in the thought of Morgenthau, Kissinger, and Niebuhr. These thinkers correctly note that the moral pursuit of ideological ends from the Crusades to recent efforts to spread democracy and free markets by force tend to backfire. Taking into account the imperfection of the world and barriers to the pursuit of the greater good through force can lead to a wise caution regarding taking up arms. This realism is not strictly amoral since it is underpinned by a sort of utilitarianism, potentially incorporating universalistic concern for non-nationals along with the citizens of the decision-making state. However, the idea that it is morally better not to pursue moral goals inevitably leads to contradictions, as it may be possible to conceive moral aims which are achievable at relatively minor costs. While realism is correct to reject a crusading moralism that verges into militarism in its violent pursuit of its ideals, it is much less plausible when it is opposed to a just war theory that incorporates principles such as just cause, last resort, and proportionality to limit the use of force. Realist insights about the difficulty of furthering goods militarily, the imperfect nature of international justice, and the obligations that states have to their own citizens,[5] make important contributions to just war thought (Coates, 2007). However, realism is implausible insofar as it implies setting aside moral conditions to pursue violence without ethical justification.

If this is right, states should and frequently do attempt to follow moral requirements in deciding when to go to war. A preventive war or intervention that will kill and jeopardize the freedom of foreign nationals cannot be justified merely in terms of national interests. Nor does an intervention's costing the interveners on balance necessarily condemn it. Just war theory must remain differentiated from realism by giving principled moral tests for the justification of war, including recognition of universal rights and obligations and not merely national concerns.[6]

In general, I take it that just war theorists of all stripes agree with the points made against realism here, such that the embrace of moral restraint against mere power politics is a shared starting point for an ethics of war. If the primary tenet of just war theory is that war is only justified under principled conditions, the next question becomes what those principles are. Although I have already introduced some *jus ad bellum* principles above, it is time to give an analysis and initial defense of them.

1.3 Just war theory and the *jus ad bellum* principles

Before I get into the criticism of new forms of just war theory in subsequent chapters, it is necessary to give background in the concepts used to discuss just recourse to war, including their moral foundations and points of contestation. Reasoned reflection and testing against our intuitions support the existence of principles regarding when the use of force is justified. A long tradition of philosophical and theological literature has sought to lay bare these principles and has been substantially incorporated in international law and politics. Of course, traditional acceptance does not prove the validity of a theory or principle. However, to the extent that rational argument, moral intuition, and social acceptance agree on principles, they are well justified. Although individual thinkers question aspects of this framework, most of the principles continue to be accepted by almost all just war theorists, such that it is a shared starting point.

The Greeks, especially Aristotle, inquired into the conditions under which war is just, separate from its effectiveness.[7] The stoics carried on this discussion of the ethical limits of violence, and the Roman Empire established legal conditions for the valid use of force (Cicero 2006). Christian reflection on the just war is usually traced to Augustine (2006), who inquired into the compatibility of war with Christian charity. He concluded that if a ruler authorizes war and does so with the right intention, aiming at peace and justice, rather than self-interested augmentation of wealth and power or anger, hatred, and lust for vengeance, then war can be justified. For Augustine, *right intention* and *legitimate authority* were the criteria for a just war (Christopher 2003, 30–48; Orend 2013, 12–3).

Augustine himself was predisposed to give great leeway to sovereigns in deciding what was for the good, permitting wars to spread the Christian faith and uphold the honor of the community and its ruler. Subsequent theologians, including Aquinas and the group known as the "schoolmen" – especially Vitoria and Suarez – sought to further refine Augustine's proposals, limiting the grounds for war. They elaborated on the notion of an objective *just cause* for war beyond subjectively right intentions, rejecting crusades against those who have done no objective wrong. They also stipulated that war should be undertaken only as a *last resort*, with non-violent alternatives attempted first. Furthermore, to be justified, the recourse to war must have a *reasonable chance of success* in achieving its ends, and there must be *proportionality* of any evil or harm caused to benefits achieved. Much the same principles became affirmed by secular enlightenment thinkers, most notably the Dutch scholar Hugo Grotius (2006), who penned the first systematic treatise on the international norms of war. Today most theorists affirm a similar list of conditions for the justice of war. Disagreements are typically about how to interpret and apply these general principles rather than whether they need replacement with some entirely different set of norms.

1.4 Last resort, proportionality, and likely success

Before getting to the questions of just cause and legitimate authority, the themes that are the focus of the book, I give preliminary analyses of the other *jus ad bellum* criteria, as they are important in their own right and will be relevant in our discussion of the authority and cause criteria. I take it that issues surrounding last resort, proportionality, and likely success are less critical insofar as recent discussions of them have been less dominated by a new permissiveness than those of just cause and right authority. At the same time, the former principles are not able to provide the same principled, bright line barrier to the waging of war that the cause and authority principles can. Nonetheless, last resort, proportionality, and likely success provide valid and important limits to the recourse to war. An understanding and rigorous application of these principles are morally necessary even when a war has a just cause and is authorized by a legitimate authority.

The last resort principle states that war should not be engaged in if the cause could be achieved by less violent means. For example, a conflict should be resolved through diplomacy instead of war if possible. How to precisely conceptualize and operationalize last resort is a thorny problem. As many commentators have noted, it cannot mean that one must try literally "everything" before war without resulting in pacifism, as more diplomacy or non-violent resistance could always be attempted. The criterion is more plausibly interpreted along the lines of "last feasible, reasonable resort," such that only those things with a likely chance of succeeding, and which do not come at too great a cost, need be tried (Coates 2007, Chapter 8; Lee 2012, 95–6). Even thus modified, it provides substantive implications restricting war-waging. The 2003 invasion of Iraq clearly violated last resort even if it was thought to have an in-principle just cause, as UN inspectors were searching and dismantling Iraq's arsenal of weapons, a process which could have feasibly and affordably been continued before launching an armed invasion (Fotion 2008, 286–7).

The criterion of proportionality states that the harm and injustice which will foreseeably be caused by war should not be disproportionate to the expected good. A war that can be expected to do more harm than good should not be fought, even if its cause is just and it is properly authorized. An otherwise just war may not be worth fighting. Importantly, one must take into account harm to the other side, so the proportionality criterion is broadly utilitarian rather than realist. There is significant debate about what harms and benefits should be allowed to count – and to what degree – in proportionality assessments. Some commentators argue that only the goods involved in the just cause can count on the plus side (Hurka 2005; McMahan 2005). Some even hold that only unjustly killed non-combatants (Lee 2012) or perhaps also combatants on the just side (McMahan 2009; Hurka 2005) are evils that count against recourse to war. Against this, I argue that all expected harms, including to any combatants, count on the negative side of the ledger for determining ethical recourse. Most people, including those who end up fighting on the wrong side due largely to coercion or ignorance, are not only not wholly evil but also not fully responsible for the threat that they pose. Their deaths, as well as obviously the deaths of those forced to fight for a just cause, are among the evils of war. During a war, killing enemy combatants may be desirable and justifiable insofar as it is a necessary means to one's (presumably just) cause, but all expected deaths count as part of the harm of war when it comes to evaluation of the validity of fighting at all.[8] If one discounts combatant deaths from the costs of war in proportionality judgments, then if entering or continuing a war like World War I did not have many expected civilian casualties, it could kill any number of combatants without becoming disproportionate. I take such inhumane and counterintuitive implications to be a *reductio ad absurdum*. Clearly, combatant casualties are a reason that war might not be worthwhile on balance and hence might be morally unjustified to enter or continue.

This discussion points to a pair of thorny problems inherent in proportionality judgments. First, parties cannot usually foresee with any precision what the results and costs of their waging war will be. During a conflict, strategies evolve, and the balance of power shifts, such that parties end up fighting a different war than the one they planned. Von Moltke's dictum that no battle plan survives encounter with the enemy seems repeatedly borne out. Another problem is that even to the extent that we expect particular results, these outcomes contain various incommensurable goods which are not easily weighed against each other (Coates 2007, 173). These include the lives of non-combatants; lives of friendly or just and unfriendly/unjust combatants; rights regarding liberty and justice short of life; injury and suffering short of death; monetary and material benefits and harms; and harms to the non-human environment and objects of cultural and aesthetic value. The goods and evils to be weighed cannot plausibly be reduced to a single objectively calculable, common denominator, whether lives, dollars, or units of pleasure.

Although the Iraq invasion is widely viewed as a failure, leading as it did to prolonged internal conflict and instability, it is possible that the state and region will eventually become more secure and rights-respecting such that the invasion could be retroactively judged proportionate. On the other hand, even the Allied fight in World War II, known widely as "the good war" with its paradigmatic case of a just cause, cannot straightforwardly be declared proportionate because its costs, including approximately 80 million killed, were so massive.

Ultimately, although we will rarely be able to say confidently that any war will be proportionate, we can recognize some wars, both past and proposed, as *disproportionate* (Orend 2013, 63). A US war to try to repel the Soviets from Hungary in 1956 or Czechoslovakia in 1968, or probably Crimea and Ukraine in 2014, would have been disproportionate. The same is true of a war with China over the status of Tibet and Taiwan and the continuation of the Korean War with the attempt to overthrow the Northern regime and reunify the Korean nation. To wage war, a party must have reason, even if imprecise, to expect that the consequences will be proportionate and be quite certain that they will not be gravely disproportionate. It would be unethical to go to war without a reasonable expectation of doing good. Overall, however, the proportionality criterion will not decisively rule out many wars and cannot be counted on to significantly restrain war on its own.

The likelihood of success criterion, like that of proportionality, follows from the logic of rational choice. It makes little sense to undertake any costly endeavor, much less one involving mass coercion and killing, when there is no likelihood of succeeding in achieving one's goal. However, a couple of problems arise. One is that if not fighting means that a large injustice will almost certainly occur, it could be rational and morally just to fight against that injustice even with slim to no chance of success. This has been illustrated with the example of the Warsaw ghetto uprising, in which Jewish concentration camp prisoners took over the camp and fought a valiant but ultimately hopeless battle against the German troops that came to recapture it. Given that they would have been annihilated anyway had they not fought, their fight was almost certainly justified. We generally consider the resistance of injustice to be inherently positive in its refusal to appease an evildoer, even if ultimately futile. On such grounds, Walzer (1977, 70–1) praises Finland's 1939 resistance to a Soviet seizure of its territory despite the Finn's inevitable defeat. For these reasons, likely success is not strictly necessary for *jus ad bellum*.

However, it must be said that resistance with little hope of success is not always worthwhile, especially if the cost of surrender would not be genocide, enslavement, or other brutal domination. For a government to order troops into battle where many will die for a losing cause is difficult to justify. Choosing to fight also typically puts non-combatant bystanders at risk as well as the combatants. While Walzer writes as if defense is always justified and perhaps obligatory, its costs make it rarely obligatory and sometimes unjustified. Indeed Walzer contradictorily acknowledges this in his later argument for states' right to remain neutral due to the risks involved in war (242–50). Perhaps one can consistently celebrate the resistance of injustice while noting a right not to do so. However, there is a point at which futile and costly resistance is no longer justified. Negotiation and surrender (or withdrawal, if on a fruitless offensive) is sometimes the best course of action.

An additional complication is that it isn't always clear what success means (Fotion and Coppieters 2008). If success means complete victory and attainment of all one's goals, few wars have a likely chance of success for either party. On the contrary, a war might achieve some beneficial results without achieving complete victory, as perhaps was true of the Finnish resistance and even the ghetto uprising. In my view, the upshot of these difficulties is that likelihood of success should be considered as an aspect of the "proportionality" of a war. The likelihood of attaining various ends clearly affects whether it is worthwhile to go to war and must be taken into account before a decision to use force is made. For example, it is clearly relevant to the validity of a contemplated armed humanitarian intervention to what extent it is likely to succeed in protecting and promoting human rights and well-being.

Last resort and proportionality (taking into account probability of success) are thus criteria that must be satisfied, or at last not clearly violated, for recourse to war to be just. We also saw that they are difficult to apply and very likely to be given conflicting assessments in the evaluation of most contemplated wars. Thus, these criteria have a limited ability to serve as a moral barrier to waging war excessively and unjustly. If war is to be constrained by decisive rational judgment, much of the burden falls upon the remaining principles of right intention, just cause, and legitimate authority.

1.5 Right intention, or subjective just cause

I mentioned above that for Augustine, fighting with the right intention was necessary for a war to be consistent with Christian charity. Over time there has been significant disagreement about what right intention involves and, correspondingly, its moral necessity. I briefly defend right intention as a just war criterion alongside just cause.

To wage war with the right intent is to subjectively attempt to achieve a just cause in one's action. Right intention cannot be simply *belief* in one's own justice, something which is probably nearly universal. Hitler's wars of aggression were not well-intended even if Hitler believed them justified. A belligerent's belief must be well-founded and reasonable. Right intention could be considered a corollary of just cause since "having" a cause usually implies that one intends to achieve it. However, it is useful to separate the issues of objective and subjective cause into different criteria. In principle, an objective just cause might exist without being the real intent of those who commit the action. For example, a humanitarian crisis might exist that would justify intervention, while a military intervention occurs without aiming to stop that crisis, as Walzer (1977, 101–5) convincingly argues was the case with the US intervention in Cuba in 1898. Or, a party might subjectively intend to right an injustice that is not actually occurring. Counterfactually, the Iraq War would have been such a case if the Bush administration sincerely and reasonably believed that Iraq had or was developing weapons of mass destruction with the intention to use them against the US. Right intent, or subjective just cause, includes a requirement that the intention is reasonable as well as sincere to rule out going to war on the basis of unsupported or negligently cultivated suspicions. The Bush Administration's manipulation of evidence suggests that the Iraq War was not even a well-intended act of preemption.

Is fighting with the right intention necessary for a war to be just? One might argue that as long as there is an objective just cause and the war serves to alleviate that cause, it does not matter what was in the mind of the war-wager. In Walzer's example, it may be that the US intervention and expulsion of the Spanish from Cuba ameliorated the humanitarian crisis without intending to do so. Similar arguments might be made regarding the invasion of Iraq if it had had positive results for the Iraqi people, regardless of whether their benefit were the intent of the US intervention. Should we not still welcome these interventions as just, while perhaps thinking less of the *character* of the interveners? In short, no, because intention partially determines the nature of an action. To kill without a defensive intent is unjustified homicide even if it turns out later that the action had positive effects (say, killing a person at random who turns out to himself or herself be a serial killer). Analogously, if a state invades a foreign country with the intent to dominate but ends up liberating it from another aggressor and failing to dominate, the attacking party, by acting with wrongful, criminal intent, was unjustified and should be condemned as an aggressor. It may be that someone could and should have invaded in order to further the just cause, and it might be *fortunate* that the unjust intervener did intervene. Yet, though its action may be welcome, the act was nonetheless immoral due to its intention. It should be noted that wrongful intentions tend to lead to wrongful actions and poor results so that the cases cited to show the intrinsic wrong of bad intent when things work out fine are largely hypothetical. In both Cuba and Iraq, the lack of humanitarian intent contributed to inhumane results.

Intent can, in principle, be distinguished from motive, a point Teson (2005), following Mill (2006), attempts to make much of. Intent, as the plan with which it is undertaken, is part of the act itself, while motive is that which gives one incentive to carry out the action. If someone rescues a drowning person with the selfish motive of a reward, the action with its intent to rescue is nonetheless praiseworthy. The rescuer evinces less than outstanding character but still acts rightly. However, if one is trying to drown a person and rescues him by accident, this culpable intention makes the action wrongful (Mill 2006, 350–1). Having introduced this distinction, Teson thinks that it justifies the invasion of Iraq. He acknowledges that the Bush Administration may have had mixed motives, including non-moral ones such as furthering Bush's political ends, avenging an assassination attempt on his father, and securing economic advantages for the US. However, Teson asserts that the invasion's "intent" was nonetheless "the liberation of the people of Iraq," a just cause (Teson 2005, 3–9). I would counter that the intention appears to have been the removal of the Hussein regime to achieve the US interests motivating the invasion, with the (partial) liberation of the Iraqi people an unintended side effect. While the motive–intention distinction makes sense in the abstract, it cannot do the work to which Teson would put it. Motives tend to affect intentions such that, in practice, it is difficult and often fallacious to distinguish the two.

It is possible to have mixed intentions as well as mixed motives, as when one is trying to accomplish several things at once or different things at different times. This relates to the question of multiple causes and the re-evaluation of war over time, which I will address in Chapters Six and Seven below. It may be nearly impossible for an intervention to be completely free of secondary intentions along with mixed motives. Since this is so, a reasonable solution is to say that right intention should predominate among intentions and should be a primary motive. One can have other motives and related goals in mind, but these should play a minor role (Coppieters and Kashnikov 2008) and should not be the basis for war entailing additional destruction and bloodshed.

The evaluation of right intention faces a particular difficulty. Because intentions are subjective and thus private, they are difficult to ascertain. This is probably why right intention is not a legal requirement for war. Parties are called on primarily to evaluate their own intentions. Nonetheless, it is sometimes possible to evaluate others' intentions. In criminal law, determinations of intent are frequently made, with intentional wrongdoing treated more seriously than unintentional wrongdoing, as in committing "murder" as opposed to "manslaughter." Agents' intentions are revealed in their actions and statements. In this manner, it is in principle possible to discern and evaluate the intentions of armed interventions.[9] To the extent that intentions are publicly opaque, individuals are still responsible for judging their own.

The last issue in the determination of right intent is that it is not necessarily clear what it means to speak of the intention of *war* as an action carried out by many agents. Various politicians and belligerents will have different goals and motives. Arguably, we should evaluate each disputant's war plans separately such that comrades in arms would be fighting different wars with different moral statuses. Yet, I suggest that it makes sense to speak of the overarching or underlying intention of a group's war, typically directed by political and military leaders. The intentions of most combatants will involve the furthering of this collective mission. A war may lack a clear mission, or the missions may be multiple. We have to look at how a war is being carried out to determine the predominant intentions. War as a whole can be judged according to these apparent aims.

The Romans formally added another criterion for *jus ad bellum* – public declaration. Although most modern writers do not include this requirement, Brian Orend has recently tied public declaration to right intention. He argues that a belligerent must openly declare its war justification, "calling its shot." This intention is the cause on which the war is to be evaluated, so it must be just. Orend requires specification of one's cause to rule out a "scattershot" offering of various shifting justifications, obscuring a war's basis and validity, as the Bush Administration did in the Iraq War (Orend 2013, 50–2).[10]

In my view, a party could have just cause and proper intention without public declaration, and it may at times be necessary to act quickly without announcement to prevent an immediate catastrophe or to gain the advantage of surprise. Nonetheless, as a general principle, a public declaration should be given. This permits deliberation about and evaluation of a proposed war. The attacking state should be open about its recourse to force with its own citizens, the party to be attacked, and the world community. Authorization processes require publicity if they are to be legitimate. In particular, if the waging of war is to be democratically accountable, citizens must know their government's military projects. Informing the target state or group that it can expect to be attacked (though not, of course, specific battle plans) may be required for last resort since it gives that party a final opportunity to come to terms without war. It also allows civilians to separate themselves from combatants, which is conducive if not necessary to satisfy the *in bello* principle of non-combatant immunity. The UN Charter requires that states explain their war intentions to the UNSC so that the international community can endorse or condemn the war plan (1945, Article 51). As I argue in Chapter Eight, legal oversight and lawfulness adherence provide rational testing, foster stability, and avoid injustice. For all of these reasons, while the declaration of war is not an absolutely necessary condition for justice, it is a moral requirement in most wars.

1.6 Just cause and the crisis of the legalist paradigm

Most everyone agrees that a just cause is necessary to justify war. To override the presumption against violence, there must be a particularly compelling reason. That reason is the justifying cause. The disputes are over how to specify what counts as a just cause. For Augustine, almost any problem perceived by a well-intentioned Christian leader justified taking up arms. Vitoria subsequently stipulated that the cause for war must be "a wrong received" (May, Rovie, and Viner 2006, 39). On this ground, Vitoria criticized the Spanish conquest of the new world in order to convert the natives to Christianity, as the Indians had committed no injustice (or rights violation) by holding their own beliefs (Christopher 2003, 53–4).

In addition to the requirement of injustice or harm done, just war thinkers have specified that the injustice must be of *sufficient magnitude* to justify recourse to lethal force on the scale of war. A modern conception incorporating these elements is that of Larry May's definition of just cause: "preventing or stopping a wrong committed by state or statelike entity ... which is sufficiently morally serious to be analogous to the risk of large loss of life that war involves" (2008, 57). A requirement of large-scale injustice means that there is a sort of proportionality judgment within just cause. However, this neither reduces just cause to proportionality nor subsumes the latter under the former, obviating one criterion or the other. There could be a just cause for war with a proportionately large injustice occurring, which nevertheless would be too costly and thus disproportionate to go to war to combat. On the other hand, a war might be proportionate insofar as it does not do more harm than good but still lack a just cause.

Until World War I, international law lacked a general conception of *jus ad bellum.* The predominant "regular war" view permitted uses of force at the discretion of heads of state (Lee 2012, 58–61). As Vattel articulated this ethic of war, "the lawfulness of its effects ... do not, externally and between mankind, depend on the justice of the cause, but on the legality of the means themselves" (Lee 2012, 61). This view essentially combines *ad bellum* realism with *in bello* just war principles. I criticized the realist view of war above. With the spread of enlightenment morality, global interdependence, and destructive military capacity, the development of international law of just recourse was probably inevitable.

The legalist, or national defense, paradigm of just cause has been traced to the 1648 peace of Westphalia, following Europe's Thirty Years War, as that conflict ended with an agreement of mutual non-interference. The legal prohibition of aggression, in conjunction with a right of self-determination, was not formally established until the Kellogg Briand pact following World War I and subsequently, and more lastingly, in the UN Charter following World War II. The Charter is founded on a conception of states as sovereign equals whose territory and self-determination are to be respected. Article 2(4) states that "All Members shall refrain in their international relations from the threat or use of force against the territorial integrity or political independence of any state." However, defending one's own or other states against aggression is permitted: "Nothing in the present Charter shall impair the inherent right of individual or collective self-defense if an armed attack occurs against a Member of the United Nations" (United Nations 1945, Article 51). As Walzer argues, actions from Hitler's invasion of Czechoslovakia, the Soviet Invasions of Hungary and Afghanistan, and the US War in Vietnam are recognizable as criminal aggression (1977, 292). The same is true of the recent US invasion of Iraq and Russian intervention in Ukraine.[11] It is noteworthy that the ICC did not define the crime of aggression until 2010. This reflects an ongoing hesitancy to make definitive judgments about just recourse in contradistinction to *in bello* acts against civilians and POWs.[12] Nonetheless, moral argument and international law concur that national defense is a just cause, and the first use of force infringing sovereignty generally is unjust aggression.

However, as discussed in the Introduction, the idea that national defense is the only just cause for war faces objections. A strict prohibition of intervention seems ethically counterintuitive. Some interpreted the Charter's Chapter VII stipulation that the UNSC can authorize force "to maintain or restore international peace and security" (United Nations 1945, Article 42) to permit humanitarian intervention. Although some would limit this to cases in which internal injustices spill across borders, many began to see human rights violations as inherently international concerns (Holzgrefe and Keohane 2003). The emergence of the R2P norm (International Commission on Intervention and State Sovereignty 2001), subsequently endorsed by the UN end as a guiding norm, further clarifies the legality of intervention to stop atrocities with international authorization. Yet, interventionist critics of legalism argue that this is still overly restrictive (Teson and Van der Vossen 2017), while some strict legalists would resist the extension (Ayoob 2002).

The Charter's stipulation of a right of defense *"if an armed attack occurs"* has been understood to rule out preventive and punitive war. Morally, preventive war seems to be barred by the requirement that an attacked party be liable due to an injustice committed. Conversely, punitive war also lacks defensive purpose insofar as the injustice to which it responds is already complete. Nonetheless, for hawks, these prohibitions are overly restrictive of the toolkit for countering threats, especially in the form of global terrorism. Interventionists argue that punitive strikes to deter repeated crimes and preventive strikes to snuff out gathering threats are required for national security and morally justifiable recourses against bad actors. There is also a concern that a legalist just cause threshold, setting a high bar for the use of force, is impractical in prohibiting the combination of several contextual considerations from justifying war. Finally, some who accept that war should only be resorted to in defensive or other exceptional circumstances defend *continuing* an already-commenced war at a lower just cause threshold. Critically responding to these recommendations for loosening the legalist just cause threshold is the subject of much of this work. I will argue on pluralist grounds that the proposed theoretical remedies are unwarranted and undesirable.

1.7 Legitimate authorization: necessary and by whom?

The situation regarding legitimate authority is somewhat different, though related. Classic just war theory held that a war must be declared by an appropriate authority. For Augustine, this entailed a recognized head of state. Legitimate authorization was necessary to distinguish war from criminal violence. The authorization requirement served to limit the proliferation of war and set its incipience in the hands of a party competent to make the decision and oriented toward making it for the common good rather than for partial, self-interested concerns. Today, objections regarding the arbitrariness of authorities, along with the proliferation of war by non-state actors, lead many just war theorists to reject the authority requirement (Fabre 2012; Lango 2014). Others who accept an authority requirement would nationalistically limit this to the belligerent's government, against Charter law's requirement of UNSC authorization (Brown 2011). I will defend the necessity of legitimate authorization to justify war and outline a process for determining legitimate authority beginning with presumptive international authorization.[13]

1.8 Conclusion

We have seen that there is a set of *jus ad bellum* principles with a long tradition of recognition, strong *prima facie* moral case, and degree of effective institutional implementation. At the same time, we have seen that there have always been doubts about moral conditions for the waging of war. Even as the just war principles are increasingly accepted, they are construed in permissive ways. Last resort, proportionality, and likely success have not been and probably cannot be understood to rule out many wars. The just cause and legitimate authority principles have the potential for being bright line norms prohibiting non-defensive and non-authorized wars, i.e. along the lines of the legalist paradigm. However, we saw that these understandings are contested, with permissive conceptions rising to challenge legalism. There are principled and practical arguments in favor of these interventionist proposals, and they are beginning to form a new orthodoxy. Nonetheless, I will argue that deontological principle, practical consequences, and institutional considerations all point toward retaining a high bar for just cause and requiring authorization procedures for recourse to war.

Notes

1 For analyses of the militaristic tendency to resort to force too easily and excessively, see Hedges (2003) and Werner (2013). I discuss these themes in more depth in Chapters Two and Five.
2 One could interpret Clausewitz differently as not asserting the ordinariness of war as an instrument, but just its political nature, in which case his view is compatible with just war thought.
3 A realist might include any harm to self-interest from the moral condemnation of others.
4 Realists might, of course, offer self-interested reasons, such as impairment to

international standing and other long-term costs, for rejecting preventive and punitive wars and other ethically objectionable measures.

5 See the discussion of the internal authorization requirement in Chapter Eight, section 8.5.

6 For a rival view about the importance of realism, see Valerie Morkevicius' *Realist Ethics: Just War Traditions as Power Politics* (2018). She illustrates that just war thinking has historically responded to concerns of power politics, and argues that the recent tradition – both Walzerian legalism and revisionist individualist cosmopolitanism – has strayed from realist insights and puts too much faith in ideal principles and international institutions. Morkevicius maintains that an injection of realism is the necessary antidote to both excessive pacifism and the exuberant crusading that she sees to be behind the war in Iraq. While it is right to note that some realists opposed that adventure and some just war theorists supported it, the just war theorists who were on board who she mentions (James Turner Johnson and Jean Elshtain) are proponents of the sort of neo-traditionalist, realist-infused just war view that Morkevicius recommends. Respect for legal and moral principles and international institutions lent strong support to the criticism of the Iraq War, while realist power politics, not moral crusading, was the driving force behind the entry into that war. In this case and in general, a principled just war theory, which takes power and consequences into account, is better than a view that rejects principle and internationalism from the start. Whether my view is excessively pacifistic, which would be Morkevicius' objection, the reader will have to decide after consulting the subsequent elaboration of just war principles and my application of them.

7 For a discussion of the historical development of just war theory, see Christopher (2003, Chaps. 1–6) and Orend (2013, 9–32). For selections from original sources see May, Rovie, and Viner (2006) or Reichberg, Syse, and Begby (2006).

8 Principles which discount concern for the unjust side are also problematic in that both sides consider themselves just and will avail themselves of the permissions given to the just side. So, encouraging the discounting of harms to the enemy causes further injustice to the innocent as well as the dis-utilitarian spread of war.

9 The reliance on action to interpret intention may make it seem that it is just action that matters and not intention. However, some actions, although not culpable in themselves, belie an underlying culpable intent, and are thus to be condemned. This shows the significance of intent as a category of moral evaluation.

10 While I agree that public declaration of a war's cause is generally a good thing, I think Orend's billiards analogy fails to capture the problem with compound justifications, as I discuss in Chapter Six.

11 Admittedly some would try to defend the legality of the Vietnam and Iraq wars.

12 We should distinguish criminal laws regarding force from the ethics of just recourse. Some acts that an ethic of war would criticize probably should not be legally punishable, insofar as the leaders may have made non-culpable errors of judgment and, even to the extent they are culpable, it may be destabilizing to make criminal charges against national leaders ordinary. In making ethical judgments about recourse to force, I do not necessarily intend those to be enforced by international criminal law. Conversely, existing limited criminal law should not exhaust our moral judgments about war.

13 See Chapter Eight below for a comprehensive discussion of these issues and defense and specification of the authority requirement.

References

Ayoob, Muhammad. 2002. "Humanitarian Intervention and State Sovereignty." *The International Journal of Human Rights*, 6 (1): 81–102. 10.1080/714003751

Augustine. 2006. "The City of God." In *The Morality of War: Classical and Contemporary Readings*, edited by Larry May, Eric Rovie, and Steven Viner, 15–20. Upper Saddle River, NJ: Prentice Hall.

Brown, Davis. 2011. "Judging the Judges: Evaluating Challenges to Proper Authority in Just War Theory." *Journal of Military Ethics* 10 (3): 133–46.

Cassese, Anthony. 2005. *International Law*, 2nd edition. Oxford: Oxford University Press.

Christopher, Paul. 2003. *The Ethics of War and Peace: An Introduction to Legal and Moral Issues*, 3rd edition. Upper Saddle River, NJ: Prentice Hall.

Cicero. 2006. "On Duties." In *The Morality of War: Classical and Contemporary Readings*, edited by Larry May, Eric Rovie, and Steve Viner, 5–7. Upper Saddle River, NJ: Prentice Hall.

Coates, A.J. 2007. *The Ethics of War*. Manchester: Manchester University Press.

Coppieters, Bruno, and Boris Kashnikov. 2008. "Right Intentions." In *Moral Constraints on War*, edited by Bruno Coppieters and Nick Fotion, 73–100. Lanham, MD: Lexington Books.

Fotion, Nick. 2008. "The War in Iraq." In *Moral Constraints on War*, edited by Bruno Coppieters and Nick Fotion. Lanham, MD: Lexington Books.

Fotion, Nick, and Bruno Coppieters. 2008. "The Likelihood of Success." In *Moral Constraints on War*, edited by Bruno Coppieters and Nick Fotion. Lanham, MD: Lexington Books.

Fabre, Cecile. 2012. *Cosmopolitan War*. Oxford: Oxford University Press.

Grotius, Hugo. 2006. "The Law of War and Peace." In *The Ethics of War*, edited by Gregory Reichberg, Henrik Syse, and Endre Begby, 385–437. Malden, MA: Blackwell.

Hedges, Michael. 2003. *War is a Force that Gives Us Meaning*. New York: Anchor Books.

Holzgrefe, J.L., and Robert Keohane. 2003. *Humanitarian Intervention: Ethical, Legal, and Political Dilemmas*. Cambridge, UK: Cambridge University Press.

Hurka, Thomas. 2005. "Proportionality in the Morality of War." *Philosophy & Public Affairs* 33 (1): 34–66.

International Commission on Intervention and State Sovereignty. 2001. "The Responsibility to Protect." http://responsibilitytoprotect.org/ICISS%20Report.pdf

Lango, John. 2014. *The Ethics of Armed Conflict*. Edinburgh: Edinburgh University Press.

Lee, Steven. 2012. *Ethics and War*. Cambridge: Cambridge University Press.

May, Larry. 2008. "The Principle of Just Cause." In *War: Essays in Political Philosophy*, edited by Larry May, 49–66. Cambridge: Cambridge University Press.

May, Larry, Eric Rovie, and Steven Viner, eds. 2006. *The Morality of War: Classical and Contemporary Readings*. Upper Saddle River, NJ: Prentice Hall.

McMahan, Jeff. 2005. "Just Cause for War." *Ethics & International Affairs* 19 (3): 1–21.

McMahan, Jeff. 2009. *Killing in War*. Oxford: Clarendon.

Mill, John Stuart. 2006. *Utilitarianism*. Reprinted in *Ethics*, 3rd edition, edited by Steven Cahn and Peter Markie. New York: Oxford University Press.

Morkevicius, Valerie. 2018. *Realist Ethics: Just War Traditions as Power Politics*. Cambridge: Cambridge University Press.

Orend, Brian. 2013. *The Morality of War*, 2nd edition. Buffalo, NY: Broadview Press.

Reichberg, Gregory, Henrik Syse, and Endre Begby, eds. 2006. *The Ethics of War: Classic and Contemporary Readings*. Malden, MA: Blackwell.

Teson, Fernando. 2005. "Ending Tyranny in Iraq." *Ethics & International Affairs* 19 (2): 1–20.

Teson, Fernando, and Bas Van der Vossen. 2017. *Debating Humanitarian Intervention*. Oxford and New York: Oxford University Press.

United Nations. 1945. *Charter of the United Nations*. Accessed November 13, 2020. https://www.un.org/en/charter-united-nations/

Walzer, Michael. 1977. *Just and Unjust Wars*. New York, NY: Basic Books.

Walzer, M. 2004. *Arguing about War*. New Haven, CT: Yale University Press.

Welch, David A. 2000. "Morality and the 'National Interest'." In *Ethics in International Affairs*, edited by Andrew Valls, 3–12. Oxford: Rowman & Littlefield Publishers.

Werner, Richard. 2013. "Just War Theory: Going to War and Collective Self-Deception." In *Routledge Handbook of Ethics and War*, edited by Fritz Allhoff, Nicholas Evans, and Adam Henschke, 35–46. New York and London: Routledge.

2 Presumptions, principles, and prerogatives in war: against hawkish neo-traditionalism

2.1 Introduction: the presumption against war and the necessity of *ad bellum* Principles

We saw in Chapter One that just war thinkers generally agree that recourse to war is only justified when a set of criteria including just cause, right intention, last resort, and proportionality are satisfied. Many also share the view, most famously articulated by James Childress, that reflection upon *jus ad bellum* begins with a presumption against war, which can only be overridden if all of those conditions are satisfied. War's distinctively large-scale lethal force requires that it meet particular justificatory hurdles before it is undertaken.

However, the presumption against war has been challenged by a hawkish strain of just war theory which seeks to re-ground just war thought in the classic theories of Augustine and Aquinas. According to this view, war is an ordinary, neutral means of pursuing justice, to be deployed by governments at their discretion, without any presumption of non-violence which must be overcome. Moreover, some neo-traditionalists argue that recourse to armed force does not necessarily need to meet all of the just war criteria, such that war becomes relatively easy to justify. For these hawks, wars from the 2003 invasion of Iraq to possible future attacks on Iran and North Korea are to be readily accepted as just. The work of James Turner Johnson (1996; 2006) and George Weigel (1987) are central to this neo-traditionalist movement which has many scholarly adherents (Cole 1999; O'Donovan 2003; Elshtain 2003; Baer and Capizzi 2005; Brown 2013).

I will begin by explaining why war is and should be thought of as presumptively unjustified. I will then respond to major arguments raised by the hawks. I conclude by discussing why all the principles are necessary for *jus ad bellum*, against conceptions of virtue ethics and the prerogatives of sovereign authorities that would waive or mitigate these conditions.

2.2 The presumption against war

Childress' (1992) case for the presumption against war (henceforth PAW) draws on Ross' concept of *prima facie* duties. Intuitively, they argue, we

DOI: 10.4324/9781003105381-2

have general, other things being equal, duties among which is the principle of non-maleficence, not harming others, along with other principles including justice, beneficence, and fidelity. Non-maleficence, Ross and Childress note, generally outweighs other duties. If we could do good, implement justice, or remain true to our word, only by harming other human beings, we generally should not do so. This is particularly true if the harm in question is the ultimate one of killing. To use a stock example, it would be wrong to kill one innocent person to harvest their organs in order to save another innocent person. Although the quantity of good done is equal to the harm (a life saved for a life lost), the duty *not* to harm outweighs the duty to do good. Similarly, the rectification of most injustices would not justify killing. Only stopping unjust killing or, arguably, other grave injustices justifies lethal force. Since war involves killing on a large scale, it is even more difficult to justify, and therefore presumptively wrong to undertake.

A PAW is implied by the accepted criteria for a just war. If war is only just if it satisfies the five or six *jus ad bellum* conditions, this implies that it is, other things being equal, something that should not be undertaken.[1] In particular, the *just cause* criterion implies that war is only justified by an especially good reason. Even more clearly, the *last resort* criterion implies that other options should be considered and, if feasible and reasonable, tried before war. These criteria both imply that war is an extraordinary measure and not an ordinary political instrument. This is to say that there is a presumption against it.

Other epistemic and practical considerations bolster the PAW. History and mathematics suggest that most wars are unjust. At most, one side in a conflict can have a just cause, for any party fighting against a just cause must itself be unjust. Moreover, sometimes neither side has a just cause, as in Walzer's (1977, 59–60) example of imperial powers fighting over who can dominate a third party. From this baseline of fewer than half of belligerents having a just cause, the number of just recourses will be fewer still, for war is also unjust if it violates right intention, proper authorization, last resort, or proportionality. It follows that in the vast majority of cases, parties in a conflict will not be justified in taking up arms.

One might think that things would be different for a generally rights-respecting liberal democratic state, perhaps with a well-intentioned, morally-versed government. Is the judgment of such a party about the justice of war not apt to be valid? Against this, I suggest that there is reason to be skeptical about even democratically elected representatives' judgments that war is warranted. There is an array of contextual and psychological reasons for the unreliability of inclinations to take up arms. Many of these are captured by the idea of a "fog of war," which undermines rational deliberation in using military force. Nationalistic passions, fear, and anger tend to combine to make one's motives for war appear more defensible than is actually warranted (Hedges 2003; Fiala 2008).[2]

Psychologically, there is a tendency to support war uncritically as fear of existential threats and excitement over communal mobilization are triggered. This enthusiasm leads parties to fail to apply or misjudge whether their contemplated wars have a just cause and are a last resort. Nationalistic passion also leads to overestimates of the ease of victory and underestimates of corresponding costs, helping to explain why victory is expected by both sides in most conflicts. As has been illustrated repeatedly in wars from Vietnam to Iraq, the intelligence establishing a war cause and likely success may be mistaken or misinterpreted. Typical deficits in human reasoning tend to be magnified in decisions about war, and national discussion tends to exacerbate rather than correct these biases. These tendencies make a mistaken judgment regarding the warrant for war particularly likely.

Recent history shows that even liberal democracies are apt to wage unjust wars. The US intervened on invalid grounds with trumped-up evidence in Vietnam and Iraq; supported oppressive regimes in El Salvador and Guatemala; and financed a brutal war to overthrow the democratically elected Sandinistas in Nicaragua. The Reagan administration's intervention in Grenada was dubious in terms of cause and almost certainly not taken as a last resort. If one includes the operations of the CIA and Special Forces, unjustified armed intervention in foreign affairs multiplies further. Lest one think that the PAW is only necessary under conservative, hawkish administrations, one should note that Clinton's Kosovo War was highly questionable in cause, last resort, and proportionality as well as notoriously lacking in UN authorization. Similar questions could be raised about the Obama Administration's intervention in Libya, which, in its assistance to rebels in overthrowing the Gaddafi regime, exceeded its UN mandate (Todorov and Johnson 2014).

In short, despite the US's and other liberal democracies' stated commitments to human rights, and possession of significant intelligence resources, they are apt to use military force misguidedly for all the reasons I just discussed. Political figures deceive the public and, sometimes, themselves in opting for military solutions. This tendency to go wrong in making decisions of such moral and practical consequence strongly implies that the burden of proof is on those who would authorize and recommend war. That is to say, that we ought to start our reflection about responses to conflict with a presumption against the recourse to war.

A final reason for the PAW is that war inevitably kills the innocent. Even if it is thought to satisfy the just war conditions, war will inevitably involve the killing, sometimes intentionally and sometimes unintentionally, of non-combatants as well as combatants who themselves do not deserve to die. Unlike a restrained use of force in personal self-defense or policing, where harm to innocents is frequently avoidable, war inevitably involves some unjust harm. This makes it an extreme and regrettable measure that should be resorted to only in exceptional circumstances.[3]

None of this is to endorse pacifism. A presumption against using armed force is compatible with a view that this presumption can be overridden in those grave and unusual circumstances in which the just war criteria are all satisfied. Indeed this is the view of Childress and other just war thinkers who understand that although sometimes justifiable, war is presumptively wrong.

2.3 The hawks' case against the presumption against war

Let us turn to the arguments of the traditionalist hawks that there is not a PAW. The most famous opponent, Johnson, develops his argument primarily on the basis of the history of the just war tradition: "The critical question is whether such a presumption is to be found in the tradition" (2006, 181). Johnson finds that the presumption against war "was nowhere to be found in the classic [just war] tradition as it took shape in the Middle Ages" (181). He continues that the tradition begins not with a presumption against war but a presumption against *injustice*. Christian love, or "caritas," gives not just a right but a responsibility to use force to protect the innocent from injustice. What Augustine and Aquinas objected to was force used with ill intent or without just cause, but not the recourse to violence *per se*. Johnson describes classic just war theory as guiding rulers in using force to achieve just ends without having to overcome any Rossian *prima facie* duty. Seconding Johnson, Darryl Charles' examination of the tradition finds that, "Just-war thinking, thus, begins not with a presumption of "peace"; indeed "peace can be cruel as Augustine well knew, but rather a presumption of justice, by which and only through which a civic peace might emerge" (Charles 2005, 340).

Johnson suggests that the idea of a PAW is a recent phenomenon, arising after witnessing the devastating effects of modern armaments in World War II. During the cold war, the risk that any war could lead to a nuclear exchange and thus global destruction created a new practical presumption against initiating belligerence. To Johnson's consternation, a presumption along the lines articulated by Childress was incorporated into a 1983 statement by the National Conference of Catholic Bishops titled "the Challenge of Peace." Johnson argues that it is a mistake to base a general theory of war's morality on contingent circumstances. He goes on to argue that at least since the 1990s, with the cold war and the risk of nuclear conflict dissipated, there is no longer any need for the PAW. He takes it that precision weaponry and humanitarian foreign policies show that well-meaning states are now disposed to use force for the greater good. Johnson suggests that the US, in particular, as a virtuous superpower, can intervene militarily for justice without significant costs. He cites examples, including the first Gulf War, of military power being used justly with benign results. In this context, we no longer need to think of war as unusually dangerous. Uses of force today, Johnson concludes, "should not be held hostage to an imagined 'presumption against war'" (1996, 36).

Another argument for permissiveness is that waging war is among the prerogatives of state sovereigns. Johnson maintains that Aquinas lists the requirement of legitimate authority first among just war criteria because the application of all the other criteria is to be made at the discretion of the sovereign. Once a just cause is determined by the sovereign to exist, last resort and proportionality are merely prudential guides about when to use force for that cause. Johnson concludes that "the presumption against war view, by reversing the weight of essential and contingent considerations, would vitiate statecraft and presume to tell sovereigns how to conduct their affairs" (1996, 34).[4] Although not strictly realists, in arguing for a Christian foundation of just war, Johnson and the other traditionalist hawks share with the realists the idea that the use of force is a matter under the authority of state sovereigns without their hands tied by international norms.

In a similar vein, George Weigel argues that just war should be thought of as "Politics, rightly ordered in truth, charity, freedom, and justice, and oriented to the pursuit of the common good" (1987, 357). The decision to use military force, on this account, is similar to the generally accepted use of police force to uphold law and order. This case against the PAW parallels a common argument against pacifism, which holds that it is contradictory to accept the everyday use of state power to keep the peace but be opposed to war for the same purpose. As Oliver O'Donovan puts it, armed conflict should be "re-conceived as an extra-ordinary extension of ordinary acts of judgment" (O'Driscoll 2008, 52). The neo-traditionalists hold that the presumption against war camp fails to ground its ethics of war in a broader account of moral values and the purpose of the state.

The suggestion is that once we think of this range of political goods and the responsibility of governments to pursue them, leaders are presumptively justified when they authorize war to pursue these ends. For most of these writers, the unique authority of governments to use violence for the greater good ultimately comes from God. This view is summarized sympathetically by Jean Elshtain, who writes, "In the Christian tradition Government is instituted by God," and its use of force is "a solemn responsibility for which there is a divine warrant" (O'Driscoll 2008, 53). There is also a secular, social contract variation of the hawkish argument for government prerogative in the use of force. On this Lockean view, governments have a responsibility to use force for the protection of the community and the preservation of natural rights within the dangerous global state of nature (Kaplan 2013, 240–6).

2.4 Reply to the traditional argument regarding the presumption against war

To the extent that their arguments rest on an interpretation of the original or essential meaning of the just war tradition, this opposition to the PAW is weak. First, other scholars have read the classic just war thinkers as presupposing that war is generally wrong even if they did not use a

Latin equivalent of the phrase "presumptively wrong" (Miller 1991). Augustine introduces the just war as "an exception to the law against killing" (2006, 15), which he thought forbade violence for even individual self-defense. Aquinas introduces his inquiry into the just war with the question "whether it is *always* sinful to wage war" (2006, 27), emphasizing war's tension with Christian neighborly love. As I argued earlier, these scholars' enumeration of a list of *jus ad bellum* criteria, including just cause and last resort, also implies that war is not generally justified.

Even if one were to conclude, with Johnson and the neo-traditionalists, that the classic thinkers did not accept a PAW, it does not follow that their view defines the tradition, such that all just war thought henceforth must do the same. When thinkers like Childress reflect on the conditions under which war is justified, they are part of the just war tradition whether or not they reach similar conclusions to its classic proponents. Johnson's attempt to reject the PAW as inauthentic is an illegitimate attempt to exclude views that differ from his own from the conversation (O'Driscoll 2008, 102–12).

Above all, it is a non-sequitur to try to settle moral and political issues with philological exposition. The central question for just war theory today is not what the earliest writers in the tradition thought, but what the correct view is. We should draw on the tradition for insight but alter earlier judgments when argument warrants it. I take it that it would be absurd to follow early just war thinkers in every instance, such as Augustine's view that any divinely authorized violence is justified, with his primary example of legitimate violence being Abraham's attempt to kill his son Isaac at God's behest (2006, 15). Augustine and Aquinas's suggestion that war is justified to make the community safe for Christianity also appears to troublingly justify wars to spread religion, arguably providing justification for wars like the Crusades and the conquest of the Native Americans. Implausible tenets of the tradition ought to be weeded out by rational argument and in light of historical experience, which includes a better understanding of the challenge of justifying armed force.

Turning to the logic of their normative arguments, Johnson and Charles draw a false dichotomy when they say that just war thought begins with a "presumption against injustice" *instead* of a PAW. Rather, as Ross and Childress sensibly argued, one has a *prima facie*, other things being equal, duty to avoid *both* injustice *and* violence. One ought to oppose and work to prevent injustice, though the duty to prevent injustice is less absolute than the duty not to commit it. Violence is not warranted to combat most injustices. For example, economic systems are rife with injustices such as exploitation, coercion, and deception. Yet, few (perhaps least of all the conservative neo-traditionalists) think that these wrongs all provide cause to take up arms. The presumption against violence usually trumps or outweighs the presumption against injustice. Only very weighty injustices overcome this presumption to justify resort to large-scale lethal force.[5]

Johnson is overly sanguine about the future of war when he suggests that the need for the PAW ended with the Cold War. War remains deadly and unpredictable. While Johnson is impressed with the ability of precision bombing to win quickly with limited immediate collateral damage to civilians, he exaggerates this as a model for war. The destruction of infrastructure in the first Gulf War and Kosovo had devastating humanitarian consequences. Limited, targeted interventions are unlikely to achieve a decisive victory and protect human rights in the long run. On the other hand, large-scale interventions seeking regime change are likely to lead to resistance and long-term instability. The mythical image of the clean war – unrealized and probably unrealizable – should in no way undermine one's general skepticism about war. The consideration of any actual war should include rigorous testing to see if it might sufficiently approach this cleanness to be justified. Moral principle and prudence both suggest that reflection about war should begin with a burden of proof to be overcome rather than a neutral attitude of broad discretion.[6]

2.5 Government authority and the presumption against war

One must also reject the hawks' attempt to derive a waiver of the norm prohibiting organized violence from the concept of sovereignty. First, hawkish proponents of sovereignty write as if governments were not circumscribed by international boundaries and international law. States are parties to international treaties, notably the UN Charter, in which they agree to refrain from using force in foreign territory. The very concept of sovereignty implies limits in its scope. If each government has a right to a monopoly on force in its territory, this implies that other states have a right against intervention (Shue 2003). To make the international use of force unproblematic, the hawks would have to explain how a particular state acquires the moral authority to enforce its sense of justice worldwide. Neither the violation of the will of God nor most injustices can be considered grounds for states to lose legitimacy, automatically warranting foreign intervention. The difficulty of legitimating international uses of force puts a burden of proof on the user.[7]

Even if sovereigns had a *legal* right to use force internationally, this delegation of power does nothing to dissolve the *moral* presumption against their use of force. Leaders ought to use their powers morally and with proper restraint. Given the seriousness of force as a measure and the ease of misusing it with disastrous results, governments do well to begin their reflections with a presumption against the recourse to violence. In themselves, moral norms regarding the use of force, including the PAW, do not deny national leaders authority to make decisions. To the extent that governments have the responsibility and right to make *ad bellum* decisions, this does not mean that every decision they make is right and should not be questioned, as Johnson implies.

The just war thinkers who would do away with the PAW as overly inhibiting the prerogative of the sovereign end up adopting what is for practical purposes a variant of amoral realism, where war is without objective moral limits. This broad discretion, with its refusal to problematize war, surrenders the critical value just war theory should give in restraining the resort to force.

2.6 Necessary principles or wise (unprincipled) judgment?

The idea of a sovereign prerogative to wage war is closely connected to a debate regarding the nature of moral and political reasoning and the status of the just war criteria. Many proponents of permissiveness regarding the use of force oppose the just war conception of a list of principles whose satisfaction is necessary for *jus ad bellum*. They argue that instead, the decision to wage war should be based on an all-things-considered judgment made in a virtuous manner by a wise leader. They find the idea of a PAW, which can only be overridden by satisfying a list of criteria, to be misguided. Johnson argues that "the classical *jus ad bellum* included three requirements: right authority, just cause, and right intention." He contrasts these "deontological criteria" with what he sees as the recently and invalidly added conditions of last resort, likely success, and proportionality. Johnson contends that the latter "are arguably prudential concerns to be taken into account in the decisions of statecraft [but] they never appeared as distinct formal requirements of the just war idea before" their introduction by Childress and other modern thinkers" (2006, 177). Johnson concludes that the expansion of requirements inappropriately ties the hands of national leaders.

A central argument against the necessity of the just war principles is that they fail to give precise and clear guidance. Critics point out conceptual ambiguity and uncertainty in the application of *jus ad bellum* concepts, including just cause, last resort, proportionality, right intention, and likely success. The principled criteria do not provide necessary and sufficient conditions and underdetermine the judgment of the justice of any particular conflict. As Chris Brown puts it, "just war thinking cannot provide us with unambiguous answers... and should not be expected to"(2013, 36) and thus, "just war thinking should not be approached as if it could provide us with an algorithm to determine what course of action to follow" (2013, 44). As a result, Brown argues for the exercise of political judgment in the form of an Aristotelian "phronesis" (or practical wisdom) made by those of experience instead of by academic and theological ethicists. O'Donovan seconds this, arguing that we cannot criticize the acts of statesmen unless "one does so from the point of view of those who performed them, i.e. without moralistic hindsight, but wars as such, like most large-scale historical phenomena, present only a giant question mark" (Fiala 2008, 25). In short, war is an art to which one cannot apply any formulaic decision criteria.

In addition to lack of clarity and precision, just war principles are charged with abstracting from reality in a way that stultifies decision-making. Baer and Cappizi object that asking whether a war meets all the criteria and overrides the PAW is an overly restrictive conception of the just war criteria. They complain that it is a "punctual conception of legal-moral decision" (Baer and Capizzi 2005, 128), which "focuses on particular provocations rather than the full political context." (127). Their argument reinforces Johnson and Weigel's view that just war theory should be conceived against the background of an account of the goal of politics in furthering the common good. Just cause, right intent, legitimate authority, last resort, and proportionality should all be understood in relation to this good and in relation to each other, instead of being assessed separately as distinct hurdles.

The limitations of abstract principles give reason to rely on judgment, dovetailing with the argument in favor of deference to sovereign authority in the last section. The neo-traditionalists recommend that the justification of war be conceived in the context of the overarching aim of government: "The right to use deadly force derives from the responsibility for the common good. Thus, government has the right to wage war to protect its common good" (Baer and Capizzi 2005, 127). This communal purpose and the sovereign's authority are in turn founded on the ultimate end: "To conceive of just war theory as a system of *prima facie* duties requires the just war theorist to relinquish his basic conviction that the forceful exercise of political power is an integral part of God's providential care for creation" (124).

Instead of moral principles, virtue ethicists found ethics on a conception of practical wisdom grounded on balanced judgment or a pursuit of the good life (Cole 1999). A general sense of justice, prudence, temperance, mercy, and charity is required beyond the knowledge of principle to act justly in concrete circumstances.[8] For the virtue hawks, rather than applying a checklist of criteria, we should think of war as an ordinary political action to be decided upon by a wise and virtuous leader, with sound judgment, dedicated to the good of the community. Such a leader will tend to pursue a just cause and take into account alternatives to and possible hazards of war but be free of absolute imperatives. Cole adds that decisions about war require "snap judgments" and that leaders need virtuous character and good judgment to make these decisions quickly without going through any systematic formula (1999, 74).[9]

For the hawks, rather than skeptically asking whether a war like the invasion of Iraq had a just cause and right intention, was truly a last resort and could be expected to be proportionate, one would ask whether the invasion fit into the overall plans for national defense and a peaceful world governed by free market trade, as determined by the presumably virtuous leadership of the United States and its allies. Last resort and proportionality would be considerations for the leaders to take into account but would not be rigorous tests that must be satisfied. Thus, most war decisions have to be accepted as just if determined just by one's political leaders.

In reply to these arguments, I would make three points. First, the just war principles are not merely prudential concerns to be taken into account but are morally required for the justification of war. It is not only just cause that is a deontological principle of right action stemming from moral reason. Proportionality and last resort are as well. One ought not to engage in violence that can be expected to do harm disproportionate to its good. To cause excessive harm due to negligent carelessness or lack of concern for the greater good is a moral evil, and its avoidance is central to fighting justly. As we saw in Chapter One, while it is possible to interpret "last resort" in a manner that makes it impossible to achieve, the underlying principle that a party should not undertake violence if its ends could be achieved by feasible less violent means follows from moral reason. Unnecessary wars, like wars without a just cause at all, should not be fought. For example, if the Iraq War was foreseeably disproportionate in consequences, or if there were measures short of war that could have been reasonably tried first, then it was wrong to go to war. We may not judge a war as harshly if these criteria are violated, as if just cause is lacking, but the action is still culpable and properly criticized.

In addition to the necessity of the principles, just war theory must defend their applicability. We can make determinations regarding the presence of just cause, right intention, legitimate authority, last resort, and proportionality. One can list wars in which each of these is fairly clearly violated as well as others in which it is satisfied, which is to say that the principles have content. One must concede that their implications are not always clear, require judgment, and can be contested. There are at least two different sorts of interpretations that need to be made. First, one must judge what each principle means. There is disagreement regarding what constitutes just cause, for example, whether preventive war and humanitarian intervention to end non-catastrophic human rights abuses are just causes. Secondly, once a conception of just cause, proportionality, last resort, etc., is agreed upon, a judgment must be made about whether the principle is satisfied by the conditions in a particular case. For example, if one had a conception of preventive war as a just cause, one would have to determine whether Iraq satisfied this cause before the 2003 invasion and, having a conception of justified humanitarian intervention, one would have to determine whether Iraq or later Libya and Syria met this threshold in terms of quality and quantity of injustices. Because this determination is contingent on a combination of factual accuracy and interpretive validity, it is certainly fallible. Nonetheless, it is not an irrational, relativistic, or intuitive process beyond the contemplation of anyone but the national leaders. A rational decision about the recourse to war will involve testing to make sure that the principles are plausibly, if not absolutely certainly, satisfied, which is to say that they are requirements for the justification of war. While advisors, scholars, and others citizens do not necessarily have all the facts and are not necessarily moral experts, they can contribute to the deliberations about whether the criteria are satisfied, contingent on their understanding of the circumstances.

As a third and final defense of necessary principles and a presumption against war, I would point out the practical advantages of an approach grounded in rational principles. The test of principled hurdles helps to rule out wars that lack sufficient justification in various ways. It provides a corrective for the confusion provided in a context of nationalism, fear, and temptation to self-interested realism. Although there may not be room for systematic principled analysis in "snap judgments," it is rare that decisions to go to war should be made in this fashion. There is almost always time for reflection, and leaders do well to take this time to think about whether a proposed act of war could satisfy the just war criteria. Individuals can develop virtues in which they are relatively good at applying the principles (sometimes without conscious analysis) and have a habit of acting accordingly. However, it makes little sense to trust esoteric judgment when the time for analysis is present.

The benefits of applying just war principles can be seen even more clearly if tested against the alternative decision-making framework presented by the virtue hawks. I will take this up in the course of replying to a final objection to the PAW.

2.7 Ethics, values, and the presumption against war

A final critique of the PAW finds its most clear expression in an article by Matthew Shadle (2012). He argues that presuming non-violence assumes a flawed moral theory in which the highest value is avoiding harm. Shadle maintains that ethics – including just war theory – must be teleological, grounded in values, in order to explain the reasons for action. For it to be wrong to kill, there must be some good, some value in life that precedes this wrong. Shadle argues, following Augustine, that "since evil is merely the absence of good, the evil is only known through the good, just as darkness is only known through light. Therefore, the general obligation to pursue human goods has a certain priority over the obligation to avoid harm" (2012, 146). The value upon which just war is based, Shadle suggests, is ultimately that of "defense of the innocent" (2012, 146) or, alternatively, realizing "justice" (2012, 148). He seconds the call of the other traditionalists for linking just war theory to a more comprehensive and substantive theory of ultimate ends. Shadle concludes that the morality of *prima facie* duties leaves obligation dubiously and dangerously ungrounded. In a similar vein, Darrell Cole complains that deontological just war lays out a list of principles "without explaining why one would want to obey these in the first place" (1999, 58), which I take to mean that the obligation lacks objective moral foundation and not just that it psychologically fails to motivate. This idea that a conception of good must undergird moral choices regarding war also points back to the idea in Johnson, Weigel, and Baer, and Capizzi that war's ethics should be considered as part of a broader view of the aim of politics.

While Shadle and the other virtue thinkers raise interesting questions about moral foundations, I find their argument unpersuasive. First, I reject the contention that morality must be based on account of the good. Secondly, I argue that regardless of moral foundations, the PAW is reasonable to infer.

The idea that morality must be teleologically founded is questionable. Although Ross' intuitive *prima facie* obligations are ungrounded, this will also be true of any moral foundation, including values such as that of innocent life. As Aristotle noted long ago, first principles necessarily lack further moral argument. This means that we have to start with axioms that are intuitively plausible but not themselves rationally demonstrable.

Perhaps a valued good makes more sense than a set of *prima facie* obligations as the foundation of ethics. However, there are problems in accounting for the ethics of killing completely in terms of values. While the value of innocent human life accounts for the justice of killing in defense, it does not account for the strong intuition that it would be wrong to kill the innocent to protect other innocents. The latter requires a deontological foundation in individual rights and duties. Shadle's qualifier of "innocence" itself presupposes deontological principles (e.g. not having done harm or wrong) and not simply a description of positive value. Shadle's alternative suggestion of "justice" as the good which guides just war thought has more potential for squaring with moral intuitions. However, it comes at the price of sacrificing clarity and substance as an ethical standard. While Shadle seems to have in mind a Platonic ideal that orients our decisions, I find this hopelessly vague if not altogether empty. Rather than any concrete good, "justice" means something like doing the right thing, appropriately taking into account conflicting claims, which puts us back in the realm of deontology. I suggest that to the extent that it has content, justice involves the balancing of several principles, including Childress' *prima facie* duties, along with some sort of Rawlsian or Habermasian procedure for determining a fair distribution of burdens and benefits.

Presumably, Shadle and other theorists of a Christian just war conceive of the ultimate justice or good as obeying the will of God. However, this good will be disputed by those with different religious views. Even for members of the same faith, it is doubtful that scripture and theology provide such clear and comprehensive moral guidance that they supersede other moral considerations, such as the presumptive obligation not to kill.

Even if the good had an ontological priority over the right, I would contend that as action guiding principles, duties such as that not to kill still have presumptive force. Any plausible view of the good, whether Platonic, Aristotelian, or Christian, will generally exclude violence. It is a strange argument that one's commitment to fundamental values disallows one from a bias against killing (as well as perhaps lying, cheating, stealing, etc.).

Following Augustine, Shadle argues that killing, unlike lying, is not intrinsically evil, as justified killing does no wrong. However, by destroying human life with its value and potential, killing is generally at odds with the ultimate good, at least as much as deception. Lethal force is paradigmatic of how humans generally ought not to treat each other. Even justified harm is still regrettable, and any degree of unnecessary harm even to the guilty is evil. Since armed force is in tension with the ultimate good, this is a good reason to presume against it. In killing, as with lying, if there are sufficient goods and justice at stake, they can override the presumption so that non-violence is not absolute.

Indeed Shadle, who shares Johnson's theological perspective, criticizes Johnson for failing to appreciate that from a Christian standpoint, one "[S]hould be predisposed toward nonviolence and only resort to violence with a sense of regret" (2012, 135 and 147–8). Shadle distinguishes this "regret" for violence from the PAW's "nonabsolute norm against the use of violence" (2012, 135). Although this distinction is intelligible, there is a strong case that the first entails the second. That which is regrettable because it is at odds with the ultimate end is presumptively wrong to undertake intentionally. Shadle seems to acknowledge this when he says that in his view, in contradistinction to Johnson, "Because of the harm caused by war, we should be predisposed toward non-violence" (2012, 135).

If one were in a context in which one had to regularly do that which were regrettable, then perhaps there would not be a practical presumption against that action. In an extremely inhospitable situation – such as a Hobbesian state of nature or a zombie apocalypse – survival might depend on eschewing the normal presumption against violence. Indeed, it is generally part of the hawks' worldview that the world is a dangerous state of nature, with the Christian or otherwise decent state surrounded by evil. Weigel's (1987) articulation of just war theory is underpinned by a Christian realist ideal of "Tranquilitas Ordinis" – a peace achieved through public order – explicitly reviving the Augustinian model. This order has to be maintained by the ready use of just force. Outside the order, it is implied, there is no morality to be followed.[10] In terms of threats to the peaceful order, the predominant evil of Communism has been replaced by Islamic terrorism. In any case, it is concluded that in this fallen world that we inhabit, for at least the time being and perhaps until the end of time, national security and nascent international order require being poised to readily use force.

However, I think this is a mistaken understanding of global security and politics. The world as we know it is not so perilous that we must set aside the PAW. There are certainly rights-abusing states and terrorist groups inclined to launch attacks against liberal democratic states, among others. But, there is not a threat of evil's imminent triumph over the good. The liberal democracies are not about to be overthrown. Statistical calculations have found the twenty-first century to be less violent than any other time in human history (Pinker 2012). In other words, there is not a supreme emergency, in Walzer's terminology, that warrants the abandonment of principled restraint. Critical reflection before resorting to violence is not only rarely fatal but also typically prudent. To accept a "barbarians at the gate" mentality in which war is an ordinary and unproblematic measure would leave powerful states steered by their leaders' sense of the good and expectations of benefits. This will inevitably result in a multiplication of unjust wars with harmful results. Just war theory should resist this devolution into what amounts to an unprincipled nationalist realism or militarism.

2.8 Conclusion

In making decisions regarding the use of force – such as intervention in Syria, strikes against Iran or North Korea, or the continuation and expansion of drone strikes around the world – we should begin with a presumption against resort to armed force. To be justified, it must plausibly satisfy the five *ad bellum* principles. We may decide that the just war conditions are satisfied and the presumption overridden in some of these cases. However, in none of them should military force be viewed as an unproblematic tool of politics. Nor is there anything in the nature of sovereign states and their rulers, moral judgment, or world politics that warrants waiving the presumption against war and the principled criteria for its waging. A just war theory that orients us toward using power justly and not excessively must begin by presuming against war.

Notes

1 Johnson (1996, 2006) and other critics of the presumption against war also sometimes argue that not all of the *ad bellum* criteria are necessary. I respond to this argument below.
2 I discuss the fog of war, self-delusion, and unintended consequences in more depth in Chapter Five on the consequences of intervention. On these topics, see also Hedges (2003) and Fiala (2008).
3 This argument that war always will involve killing the innocent is central to pacifistic arguments, such as May's (2015). In my view, the fact that we foresee that injustice will be done in the course of war does not absolutely rule out undertaking war to prevent greater evils in those cases where the just war criteria are satisfied. Such arguments bolster a presumption against war but not a categorical rejection of its use (even for the present era, as I take May's contingent pacifism to hold.)
4 The reference to preference for "contingent" matters has to do with the argument that the presumption against war view stems from an assessment of the current dangers of war, a contingent matter.
5 Some opponents of the presumption against war argue more consistently that the just war criteria are not all necessary conditions for war. I respond to this view in Section 2.6 below.
6 One might wonder whether some uses of force short of full-scale war are sufficiently minor that the presumption against war does not arise. I take this up during my discussion of the consequences of war for human rights in Chapter Five.

7 I discuss the bar to intervention posed by sovereignty in Chapter Four. In Chapter Eight, I defend an internationalist conception of the legitimate authorization of war.
8 Cole (1999) argues that specifically the theological virtues are required for moral action.
9 Virtue ethics is not always interpreted as having hawkish implications. David Chan (2012), while sharing the views that just war principles do not provide clear direction and that decisions should be made according to broadly wise practical judgment, concludes that a virtuous leader would only resort to war in exceptional circumstances.
10 This view is moderated in part by Weigel's suggestion that *tranquilitas ordinis* must itself be rooted in caritas – love of neighbor – as well as power. Weigel also holds the possibility of global *tranquilitas ordinis*, so that his vision is not strictly nationalist. Nonetheless, he holds that for now, the concept justifies the ready recourse to war by ordered societies threatened by forces of disorder.

References

Augustine. 2006. "The City of God." In *The Morality of War: Classical and Contemporary Readings*, edited by Larry May, Eric Rovie, and Steven Viner, 15–20. Upper Saddle River, NJ: Prentice Hall.

Aquinas, Thomas. 2006. "*Summa Theologica*." In *The Morality of War: Classical and Contemporary Readings*, edited by Larry May, Eric Rovie, and Steven Viner. Upper Saddle River, NJ: Prentice Hall.

Baer, Helmut, and Joseph, Capizzi. 2005. "Just War Theories Reconsidered: Problems with Prima Facie Duties and the Need for a Political Ethic." *Journal of Religious Ethics* 33 (1): 119–37.

Brown, Chris. 2013. "Just War and Political Judgment." In *Just War: Authority, Tradition, and Practice*, edited by John Williams, Cian O'Driscoll, and Anthony Lang, 35–48. Washington, D.C.: Georgetown University Press.

Chan, David. 2012. *Beyond Just War: A Virtue Ethics Approach*. New York: Palgrave McMillan.

Charles, J. Darryl. 2005. "Presumption Against War or Presumption Against Injustice? The Just War Tradition Reconsidered." *Journal of Church and State* 47 (2): 335–59.

Childress, James. 1992. *Moral Responsibility in Conflicts*. Baton Rouge: Louisiana State University Press.

Cole, Darrell. 1999. "Thomas Aquinas on Virtuous Warfare." *Journal of Religious Ethics* 27 (1): 57–80.

Elshtain, Jean Bethke. 2003. *Just War Against Terror: The Burden of American Power in a Violent World*. New York: Basic Books.

Fiala, Andrew. 2008.*The Just War Myth: The Moral Illusions of War*. Lanham, Maryland: Rowman & Littlefield.

Hedges, Michael. 2003. *War is a Force that Gives Us Meaning*. New York: Anchor Books.

Johnson, James Turner. 1996. "The Broken Tradition." *The National Interest* 45: 27–37.

Johnson, James Turner. 2006. "The Just War Idea: The State of the Question." *Social Philosophy and Policy Foundation* 23 (1). 10.1017/S0265052506060079

Kaplan, Shawn. 2013. "Punitive Warfare, Counterterrorism and *Jus Ad Bellum*." In *Routledge Handbook of Ethics and War*, edited by Fritz Alhoff, Nicholas Evans, and Adam Henschke, 236–49. New York, NY: Routledge.

May, Larry. 2015. *Contingent Pacifism*. Cambridge: Cambridge University Press.

Miller, Richard. 1991. *Interpretations of Conflict: Ethics, Pacifism, and the Just-War Tradition*. Chicago: University of Chicago Press.

O'Donovan, Oliver. 2003. *Just War Revisited*. Cambridge: Cambridge University Press.

O'Driscoll, Cian. 2008. *The Renegotiation of the Just War Tradition*. New York: Palgrave MacMillan.

Pinker, Steven. 2012. *The Better Angels of our Nature*. London: Penguin.

Ross, William David. 2012. "The Right and the Good." In *Ethics*, 5th edition, edited by Steven Cahn and Peter Markie, 475–84. New York: Oxford University Press.

Shadle, Matthew. 2012."What Is at Stake in the Debate over Presumptions in the Just War Tradition." *Journal of the Society of Christian Ethics* 32 (2): 133–152.

Shue, Henry. 2003. "Limiting Sovereignty." In *Humanitarian Intervention and International Relations*, edited by Jennifer Welsh, 11–28. Oxford: Oxford University Press.

Todorov, Tzvetan, and Kathleen Johnson. 2014. "The Responsibility to Protect and the War in Libya." In *The Ethics of Armed Humanitarian Intervention*, edited by Don Scheid, 46–58. Cambridge: Cambridge University Press.

Walzer, Michael. 1977. *Just and Unjust Wars*. New York: Basic Books.

Weigel, George. 1987. *Tranquilitas Ordinis: The Present Failure and Future Promise of American Catholic Thought on War and Peace*. New York: Oxford University Press.

3 Why punishment is not a just cause for war

3.1 Punitive war: introduction

A common defense of the US wars in both Afghanistan and Iraq is that they were justifiable punishment: the former for the 9/11 terror attacks and the latter for the Hussein regime's use of chemical weapons against the Kurds and the assassination attempt on George H.W. Bush. In the early just war tradition, punishment was considered a just cause alongside defense. Augustine conceived of punishment as war's principal justifying purpose:

> As a rule just wars are defined as those which avenge injuries, if some nation or state against whom one is waging a war has neglected to punish a wrong committed by its citizens, or to return something that was wrongfully taken. (2006, 82).

Sovereigns were understood to have a God-given right to enact justice universally. Even modern thinkers, including Grotius and Locke, held that states could justly fight to punish rights violations beyond their borders if there was no other recourse for the aggrieved (Kaplan 2013, 240–3).[1] Recent just war thought has tended to reject this view, holding that only defending against injustice in the form of aggression or human rights abuses can justify war. This view is reflected in UN Charter law, which recognizes defense against an armed attack but not punishment as a just cause. Due to jurisdictional concerns about punitive force beyond one's own sovereign territory as well as due to the destructiveness of war combined with the limited value of retribution, punishment of injustice after the fact came to be widely rejected as a just cause.

However, there has been a recent revival of just war thought arguing that punishment can be a cause for war. For neo-traditionalists like James Turner Johnson, punitive war follows from the Augustinian view that it is the duty of the sovereign state to act anywhere to uphold the *tranquilitas ordinis*, a lawfully organized peace. While I criticized this broad discretion for states and their leaders in the last chapter, the idea of punitive war warrants its own discussion. Several recent thinkers have argued, without

DOI: 10.4324/9781003105381-3

relying on dogmatic religious or other traditionalist conceptions of sovereignty, that punishment should be recognized as just cause for war. In their view, the legalist rejection of punitive war fails to allow for adequate security measures in response to threats short of a full-scale invasion, including especially terror attacks. In the context of a lack of consistent international law enforcement, punitive war theorists argue that states need to engage in self-help in response to attacks. Proponents add that many widely accepted uses of force have been punitive and not merely defensive.

I will argue that this is a mistake, that punishment is not a just cause for war, and that a just war must be defensive. I begin by discussing the nature of punitive war and the reasons for its recent rejection. I then introduce what I take to be the strongest defense of punitive war before rejecting it on deontological and consequentialist grounds. I evaluate and interpret suggested punitive wars in light of my moral and conceptual framework. I finish by distinguishing between punitive war and reprisals and law enforcement measures short of war, which can be justified punitively in certain circumstances.

3.2 Working definitions of war, just cause, and punitive war

I begin by defining key terms. By "war," I mean a large-scale violent conflict between groups struggling over political control (Orend 2013, 3). I take it that there are at least two criteria for a just cause for war. First, those to be attacked must be responsible for a sufficiently grave injustice to make themselves liable to lethal force. Second, the war to be fought must have a sufficiently important purpose, typically gains in human rights enjoyment, to justify the massive harms and human rights infringements, including the killing of the innocent, which are inevitably involved in war (May 2008; Lee 2012; Orend 2013).

By 'punitive war,' I mean a war that responds to an injustice already completed without trying to stop or defend against that injustice and which is enacted according to a conception of punitive justice rather than mere vengeful retaliation. Punitive justice can take two broad forms. First, it can be retributive, aiming to give deserved punishment to wrongdoers.[2] Or it can be utilitarian, seeking to achieve deterrence or other beneficial consequences, including incapacitation or rehabilitation, through punitive force.[3] Deterrence can be either "general" (aimed at dissuading any actors from aggression and rights violations by the threat of receiving a similar punishment) or "specific" (dissuading the same criminal party from repeating this crime or similar injustices). Despite their moral rationale being future-oriented, utilitarian aims can still be punitive because insofar as they follow a rule of responding to past injustice. In the next sections, I take up these forms of punishment, in turn, to show why neither is independently a just cause for war.

3.3 Retribution and war

A war of retribution can potentially meet the first requirement of just cause insofar as it involves targeting a party that has committed an injustice that may make it liable to attack. However, retribution almost inevitably fails the second requirement. Even if the infliction of proportionate harm on a wrongdoer is of moral value, as the concept of retribution presupposes, this value is not sufficient to justify violence that will almost certainly cause large amounts of harm to parties who bear little or no responsibility for the initial injustice. As others have commented, war is an indiscriminate tool for administering punishment (McMahan 2008, 78–84; Luban 2011). The political and military leaders primarily responsible for aggression are apt to avoid fighting and escape capture, while innocent non-combatants and relatively innocent combatants are killed in large numbers. If the US's wars in Afghanistan and Iraq were carried out only or mainly for the sake of retribution, they were monstrously disproportionate. While the Nuremberg Trials were probably justified after World War II, getting to conduct those trials was not and could not have justly served as the Allied purpose in fighting the war. Retributive war almost inevitably violates the second, proportionality component of just cause. Moreover, insofar as it is indiscriminate, a retributive war even fails to consistently satisfy the first criterion, liability.

None of this is to say that retribution cannot be undertaken against war criminals and human rights violators. They can be tried for their crimes – either after defeating them in a war with an independent defensive purpose (as in the Nuremberg case) or through their fortuitous apprehension without a war. However, administering criminal prosecution and punitive sanctions cannot provide a war's justifying cause.

3.4 Deterrence and other benefits as causes

If retribution cannot be a just cause, what about deterrence and other benefits of punishment? Preventing future aggression, including terror attacks, might have sufficient value to make a war proportionate. Defenders of punitive war argue that a forceful response can deter aggression and that without a punitive response, aggressors will be emboldened. As Kenneth Kemp puts it, "the government has the duty and the right to protect the common good of the community it governs," and "punishment through its preventative and deterrent effects, is one of the ways a nation has of defending itself against external threats" (Kemp 1996, 343–4). Punishment has also been argued to achieve other goods, including showing symbolic respect for the law and the rights of victims (Feinberg 2014), paradigmatically in war crimes trials such as those held by the ICC at the Hague as well as those at Nuremberg. Since it is good and perhaps obligatory to respect human rights, and punitive war does this, it can be concluded that it is good and perhaps obligatory to conduct punitive war.

However, valuable consequences alone cannot justify punitive war for the reasons that non-consequentialists have long objected to utilitarian theories of punishment. Deterrence of future injustice and positive symbolic expression cannot themselves make a party that is not responsible for unjust harm liable to lethal attack.[4] If the Afghanistan and Iraq interventions were undertaken for the sake of demonstrating the United States' willingness to use force to third parties, these wars were wildly indiscriminate. Consequentialist punishment cannot satisfy the first criterion of just cause for war. Numerous theorists, prominently including Kant, Hart, and Rawls, have argued that good consequences are not sufficient to justify punitive harm to an individual, especially capital punishment (Feinberg et al. 2014, 691–6). To the extent that punishment for deterrence and other benefits is intuitively justified, it is because we tacitly assume the responsibility of the party to be punished. Indeed, *specific* deterrence, incapacitation, and rehabilitation seem more plausible as rationales for punishment than general deterrence, as the former exercise coercion to prevent further crimes by the coerced agent rather than by society or the world in general. To the extent that utilitarian rationales contribute to the justification of punishment, it is only in combination with deontological responsibility, which creates a liability to coercion if not desert of retribution.

3.5 Justifying punitive war with a combination of retribution and benefits

Since neither of the main rationales for punishment is sufficient to justify war, an obvious suggestion, and I think the only potentially viable justification, is a compound one. Retributive considerations could make a party liable to attack, while the deterrent and other benefits make the military response to that party sufficiently worthwhile to justify the destructiveness of war. Indeed, such a combined justification has been offered by new proponents of punitive war. Shawn Kaplan offers a defense of punitive war contingent on the level of threat and expected consequences. He draws on Locke's and Grotius' arguments that rights violations can be punished across borders if there is no higher authority to appeal to. Locke and followers such as Christopher Wellman argue that rights violators forfeit their right not to be punitively struck and thus are liable to attack (Wellman 2012). Moreover, if terrorists and other rights violators could inflict harm with impunity, this would seem to invite further attacks. A punitive response against foreign rights violators may serve to deter future attacks or incapacitate the attacking party. For these reasons, Grotius defends punitive war to protect rights and further international peace. Kaplan agrees with Grotius that separate from rights forfeiture, states have an obligation, and thus a right, to punish if required to preserve their own citizens' rights:

the victim state is viewed as having the authority to punish the cross-border militants or terrorists due to its obligations to protect citizens' basic rights from such attacks. If deterrent retaliation is a necessary means for preserving these rights, then the state's fundamental obligation to preserve its citizens' basic rights would establish the moral permission (i.e. authority) to carry out the punitive warfare. (Kaplan 2013, 244)

States harboring terrorist groups, as well as the terrorists themselves, might be liable to punishment, though Kaplan notes that a state's negligent failure to stop terrorism is more dubious as a just cause than directly planning or materially assisting attacks. Because the punitive authority derives in part from a state's responsibilities for protecting its own citizens, Kaplan restricts the right of punishment to the victim state (or group).

I take it that in this argument, criminal culpability – and the corresponding vindication of rights – and utility combine to justify punitive counterterrorist strikes. This compound justification parallels what some view as the best general theory of punitive justice, one which combines consequentialist goals with a limiting condition of proportionate retribution (Bedau and Kelly 2019). Such a rationale could justify the US's punitive strikes at the Al Qaeda perpetrators of the 9/11 attacks in Afghanistan, as well as perhaps the overthrow of the harboring Taliban.

Thomas Hurka offers a similar justification for punitive war, at least as a follow-up to defensive war. He argues that only a massive injustice such as aggression or a humanitarian catastrophe can make a state or other group liable to attack. However, he adds that this liability is "general," such that force can be used against the criminal aggressor for additional purposes beyond defending against its unjust attacks. Among his proposed additional "conditional" war causes, Hurka specifically includes the punitive measures of "forcibly disarming an aggressor, deterring future aggression, and preventing humanitarian crimes (by rehabilitating the target state)." These purposes of incapacitation, deterrence, and rehabilitation correspond to the major utilitarian aims of punishment. Hurka cautions that "if one only has the conditional cause, one is not permitted to fight." However, "once another, independent just cause [typically defense against aggression] is present, a conditional cause can become a legitimate goal of war" (2007, 201). Hurka cites examples of each of the three types of conditional, punitive just cause. Deterrent aims, he claims, justified Britain's attack on Argentina over the Falklands. The allied disarmament of Germany and Japan at the end of World War II was justified incapacitation. And, the then ongoing US intervention in Afghanistan was rehabilitative. Each of these punitive purposes, with their beneficial consequences, is seen as justified because it follows an act of aggression that made the party liable to war measures (Hurka 2007).[5]

In terms of punishment theory, like Kaplan, Hurka does not explicitly call his theory of just cause a hybrid of deontological retributivism and consequentialism. However, it is clear that it is, as utilitarian benefits justify punishment in combination with a limiting condition of liability based on individual responsibility: "punishment still requires liability, since only a criminal who has made himself liable by criminal action may be punished. But the liability is only the more global one of having acted criminally rather than any specific liability" (Hurka 2007, 204).

Kemp (1996) gives several examples of what he takes to be justified and widely accepted punitive wars. These include the US chasing Pancho Villa into and around Mexico between 1914 and 1917, following his raid on a US town; the 1968 Israeli destruction of 14 planes at the Beirut airport in response to Lebanon's support of P.L.O. terrorists; and the 1993 US bombing of Iraq's Mukhabarat intelligence headquarters following the assassination attempt on former president Bush.[6] These cases are meant to show the intuitive validity of punitive war and add to Kemp's argument from the sovereign's duty to protect. Anthony Lang suggests additional punitive interventions, including by the US in Panama to apprehend Manuel Noriega (1989–90) and against Libya in 1986 in response to terrorism, and 1982 and 1996 Israeli incursions into Lebanon (Lang 2005, 50).[7] One might add to this list the 2017 US strike on Syria's Shayrat Airbase in response to a chemical weapons attack allegedly launched from there. I take it that insofar as these are examples of justified punitive war, it is on the compound ground discussed here. The actions focused on responsible parties[8] and seemed to be aimed at deterring similar future attacks, a potentially proportionate consequence.

3.6 Response to punitive war on mixed grounds: the authority problem

I argue that punitive war *cannot* generally fulfill the liability criterion or the value criterion and thus should be rejected. A first problem is that liability to punishment generally requires a finding by a legitimate authority conducting a fair procedure. Such due process is typically not accorded in warfare, making it dubious as a means of punishment. Moreover, punitive strikes are typically carried out by the aggrieved party. It is a common principle of justice that parties should not be judges in their own cases. Biases and desires for vengeance lead to excessive punishment. To the extent that punitive war is apt to punish the wrong parties or punish excessively, this is an additional reason to reject it as a just cause (Luban 2011, 317–18). Punitive war is typically implemented by parties acting outside of their jurisdiction on a group that has not agreed to be governed by its authority.[9]

Defenders of punitive war suggest that internationally, parties are in a state of nature in which communities – of necessity – can justly defend their own rights. Nonetheless, there is existent international law with agreed-upon norms and an institutional framework for their interpretation, including the judgments of the ICC, ICJ, UNSC, and UN General Assembly. The rules have been consented to and are widely recognized as normative (Claude 1966). International actors defend the legality of their actions and criticize others for violating international law (Franck 1990). While few legal violations receive UNSC – authorized military punishment, other sanctions, as well as loss of international stature and economic opportunities, amount to a substantial enforcement regime. I discuss legitimate authority in more detail in Chapter Eight. For now, I suppose that these bodies do constitute law and that there is an obligation, although not absolute, to generally defer to their decisions regarding uses of force that do not defend against an ongoing or imminent attack. The existence of an international legal and security system means that states cannot exercise unilateral punishment as if they were in a state of nature.

However, this argument alone is not decisive against punitive war. On the one hand, a failure of international authorities to enforce international norms might at times provide a right of states to act unilaterally, although they would have the burden of proof. On the other hand, the Security Council itself might approve punitive war, such that legal procedure is observed and the legitimacy objection does not apply.[10] The argument here casts doubt on unilateral punitive war (which is currently the most frequent form) but not internationally authorized ones.

3.7 Questions about liability to punitive war

I have discussed as a criterion of just cause that those attacked must be liable to lethal force. This parallels the criminal justice principle that only the guilty should be punished. However, war attacks are not typically targeted according to individual responsibility. War involves a large-scale struggle in which any members of the opposing military are taken as targets, and the collateral killing of non-combatants is accepted according to the doctrine of double effect. A punitive war that seeks to achieve deterrence and incapacitation alongside retribution will be especially apt to harm the innocent. The logic of deterrence recommends visibly destructive results that cripple the military forces and infrastructure of the group attacked, almost inevitably harming non-combatants along with military forces. The worry here is that punitive war, by needing to achieve results along with measured justice, will tend to sacrifice the latter. Recall Donald Rumsfeld's post 9/11 lament that "there aren't any good targets in Afghanistan and there are lots of good targets in Iraq" (Leung 2004), suggesting a strategy of attacking where American military might could be displayed more clearly. The history of domestic punishment also shows the tendency for concern for deterrence to lead to brutally inflicting harm beyond that to which individuals are liable as in measures such as public floggings and executions. The domestic penal code has developed rules to avoid excessive force beyond an individual's liability. War has not had and is not likely to be able to incorporate rules preventing force beyond liability.

A related problem is that war is conducted against groups, either state or non-state, while moral responsibility accrues to individuals and is thus variable within the group to be punished. It is true that we speak of states and organizations, like corporations, being responsible for actions as a collective, for example, for paying financial debts and compensating injured parties. Collective responsibility can be a useful fiction so long as it does not violate the basic rights of individuals. Economic sanctions, dispersed through taxes to be shared by the whole society, place a relatively small burden on innocent parties. Admittedly, sanctions themselves can take lives if they lead to shortages in food or other necessities. Indiscriminately lethal sanctions are themselves an unjustifiable punishment. In any case, the administration of lethal force on the grounds of collective responsibility is a clear violation of human rights. Punitive war inevitably will kill the innocent, contravening the liability principle central to both just cause and punitive justice.

It is theoretically possible for a punitive war to target only those determined responsible for crimes, typically the political leadership and perhaps, the military forces that carry out their aggression and defend them from capture and punishment. If enemy combatants pose a lethal threat while defending their (presumably culpable) leaders, this can make them liable to attack at that point. However, in a problem shared with preventive war, an initial punitive strike targets those who are not currently engaged in any harm or threat and thus who are presumably not liable to attack. In this regard, punitive war is even more problematic than a war that seeks to prevent some threatened future harm.

One might object that the combatants in the group to be punished may have themselves committed crimes or should at least be aware that they are fighting for a group that has engaged in such crimes. If so, they might either be liable to punitive attack themselves or liable for trying to prevent the justified punishment of their leaders and comrades. However, many fighters and civilians will be unaware of their group's aggression and other crimes and may be excused for being prepared to defend their nation or group against an attack. They may not have heard or been able to evaluate the attacking party's claim that it acts with punitive justice. Insofar as these combatants are fighting against what is justifiable punishment, posing a threat to the punitive forces, they could become liable to attack. However, it could not be the goal of a just punitive war to inflict harm on such individuals. Drawing them into a conflict where they become liable to be killed would be a negative side effect, counting against the proportionality of such a fight. The constraint of limiting attacks to those who are liable to punishment means that most punitive strikes would have to be surgically targeted at those culpable or responsive to threats that occur in the course of the mission.

There is some question about the justification of lethal attacks against even the leaders responsible for criminal aggression when those individuals currently pose no threat. At least three conditions are assumed in justifying the extreme measure of extra-judicial killing. First, their crimes must be large enough to make them liable for the level of force administered, especially if it is to be lethal. Second, punishment should be determined by a legitimate procedure. Individuals should not be targeted with lethal force without evidence and due process. Third, punitive force should seek to capture individuals believed guilty of crimes rather than kill them, if possible. Just punishment, as opposed to vengeance, seeks to administer force in a measured and proportionate manner. These limits conditions imply that punitive force, unlike direct defense, must avoid the broad and imprecise force that characterizes war.

We saw that Hurka's "general liability" is an attempt to explain why aggressors can be attacked for new purposes beyond defense. However, there are several problems with this idea. One is that in treating states as groups liable to ongoing non-defensive attack, it permits the use of force against individuals who have no responsibility for injustice and do not actively pose a threat. The goals of deterrence and incapacitation of attacks in the future fail to exercise discrimination between criminals and combatants on the one hand and non-criminals and non-combatants, who might eventually pose a threat or be useful to eliminate, on the other. Even if we accept that the liability to attack diffuses throughout a criminal state or group, or at least its armed forces, a second problem arises. Punitive war, guided by an attempt to achieve long-term deterrence, incapacitation, and rehabilitation, will tend to cause destruction out of proportion to the harmed individuals' particular liability.

In terms of the ethics of punishment, in domestic law, individuals cannot be given additional punishment for the sake of desirable ends beyond that to which their criminal responsibility makes them liable. Deterrent and rehabilitative benefits are welcome but cannot justify additional quantities or qualitative severity of punishment. Hurka's defense of punitive war aims does not stipulate limits on the force to which a former aggressor is liable, suggesting unlimited liability at odds with justice. Of course, it would be open to Hurka or others to stipulate that the total punishment administered to a group (or more challengingly, to each individual) militarily would be limited by a liability quantity. However, in practice, it is impossible to keep punishment measured in war, so the concept of general liability would remain an invitation to unlimited deterrent, incapacitating, and reconstructive force. Unlike domestic law's precise sentence assigned to individuals according to guidelines for their particular crimes, a war will tend to exceed any liability requirement as it strives to achieve deterrence, incapacitation, or rehabilitation. For example, the open-ended war on terror, and the long US intervention in Afghanistan in particular, appear to lack specified purposes and limits, such that it would be difficult to begin assessing these conflicts according to any standard of punitive justice.

In order to meet the first, liability criterion of just cause, punitive force would have to take the form of small-scale targeted reprisal strikes or law enforcement actions short of war. This might include raids to arrest or kill specific individuals, such as that which killed Osama bin Laden. It also could include reprisals or law enforcement strikes aimed at arms or military capacity. I discuss the ethical limits of reprisals and law enforcement measures as punitive alternatives short of war in section 11 below. For now, I point out that the seeming justifiability of limited punitive measures short of war does not undermine my thesis condemning punitive war.

Kemp denies that there is a morally significant difference between small reprisals and law enforcement measures short of war and punitive *wars*. He suggests that the justificatory logic is the same: "It is violence that triggers the moral questions and a single act triggers those questions just as completely as does a world war" (Kemp 1996, 337). I think this is wrong. In entering a war, with its widespread violence and struggle over the governance between communities, there is much more at stake. Thus, war warrants a higher threshold of justification. Additionally, unlike law enforcement, war permits and generally involves the free targeting of enemy combatants and civilian collateral damage. As a result, war inevitably kills the innocent and is thus inherently problematic as a means of implementing justice. These are reasons for a high threshold of just cause for war and a reason to think that war is not well suited to punitive justice.

Nonetheless, arguably, in a rare instance, it could be known that either responsibility for aggression spreads widely through the armed forces or that in order to punish the responsible parties, it will be necessary to fight armed forces who are culpably prepared to defend them. In the case of a terrorist group like Al Qaeda or Isis, one might argue that all those taking up arms for their cause are already involved in a murderous plot and thus liable to attack. In such a case, war could be argued to satisfy the liability condition of just cause. We must now consider the possibility of wars satisfying the second, proportionality component of just cause.

3.8 The limited utility of punitive war

Earlier, I raised the possibility that deterrence and other benefits of punishment could justify the harm done by a punitive war. However, I will now argue that in practice, punishment cannot be expected with confidence to have such benefits to a sufficient degree to justify war. The attempt to send a lasting, decisive message through armed force is likely to be disappointed. While the outcomes of any war, including one of immediate self-defense, are uncertain, the value of deterrence is especially doubtful due to its remoteness and complex material and psychological contingencies.

Let us begin with strategy of "general deterrence," meant to warn other parties not to undertake similar aggression for fear of similar punishment. International punishment is inevitably inconsistent due to considerations of relative power, interests, resources, and political will. This means that past punishment will not be taken as predictive by future potential aggressors, undermining deterrence (Walzer 1977). To take recent examples, the US invasions of Afghanistan and Iraq do not seem to have had a substantial deterrent effect on terrorism, aggression, chemical weapons use, or assassination. Parties considering such actions may correctly infer that the US is unlikely to launch a third such punitive invasion given the costly nature of its recent conflicts. Isis and Al Qaeda affiliates, as well as Iran and North Korea, appeared to be emboldened rather than cowed by the Iraq intervention. Non-state actors who launch terror strikes or commit other international crimes may hope that they will not be traceable or considered geopolitically not worth pursuing. Aggressors may also be undeterrable due to a passionate commitment to their cause fueled by anger, resentment, and hatred, as well as religious or other ideology. Moreover, individual military and political decision-makers stand to benefit from escalating conflict regardless of whether this is in the interest of the group they represent. Attacks on one's group tend to lend support to hard-liners seen as champions of the nation, such that provoking attacks becomes a rational strategy for militant political leaders. Thus, punitive strikes can have the opposite effect of deterrence, leading to a cycle of retaliatory violence. These limits, contingencies, and uncertainties of general deterrence make it of dubious value and unlikely to counterbalance the harms inherent in war as a means.

Specific deterrence and incapacitation of an aggressor appear to be more reliable benefits than general deterrence. A state or other group which suffers punitive force seems particularly likely to take a lesson that it may be attacked again if it continues its aggression. In the Kemp examples, I take it that the main message and expected benefit was specifically the cessation of attacks by Villa, Iraq, and the PLO through Lebanon, respectively. At the same time, punitive strikes can have an incapacitating effect on the opponent. "Lessen[ing] their capabilities" is the goal of punitive measures defended by Kaplan (2013, 245). Indeed, attrition seems to be part of the counter-terrorism plan seen in US drone attacks. Unlike a dubious preventive war against a gathering threat, a punitive deterrent and incapacitating war responds to an already committed wrongdoing, increasing the need for prevention as well as the liability of the party to be attacked.

Nonetheless, expectations of specific deterrence and incapacitation, like general deterrence, are frequently exaggerated. Rational strategy and passionate commitment will make aggressors, particularly sub-state groups who have taken up arms for a cause against a more powerful force, unlikely to back down after suffering a punitive attack. Punitive attacks are likely to invite further reprisals and inspire new enemy recruits, negating attrition. For example, it is unclear whether decades of US and Israeli wars against terror groups have had a significant deterrent or incapacitating effect. It is also clear that the invasions of Afghanistan and Iraq have served as recruiting points for terrorist groups. It seems likely that the incitement effect of the ongoing interventions has equaled or outweighed deterrent and in-capacitation benefits, making these disproportionate as punitive wars.

As with preventive war, since there is no ongoing act of aggression, there is time to try non-violent measures, first, and reprisals short of war, second, before resorting to punitive war. Whenever the preventive value of a puni-tive war could be achieved by other means, the attack should be rejected as unnecessary. The fact that punitive war (like preventive war) typically fails the last resort criterion is another reason that punitive war should not be viewed as a just cause. Indeed, after defending punitive war in principle, Kaplan (2013, 245–6) notes that punitive war could have these sorts of negative effects, which might, in turn, outweigh its deterrent and in-capacitating effects, and making punitive war disproportionate and unlikely to succeed. If just causes must have rule utility, these consequential problems tell against punishment as a war rationale.

Overall, the uncertain benefits and certain direct harms and likely side-effects of punitive war mean that it will generally fail the second criterion for just cause. As it is not the sort of thing that can be expected to be beneficially pursued by war, punishment should not be viewed as a just cause. Since its principled justice in hitting liable targets was also dubious, these con-sequentialist considerations combine with the deontological weakness to condemn punitive war.

3.9 Examples of justified punitive wars?

What of the cases of arguably justifiable punitive wars cited by Kemp and others? I argue that insofar as these were justified, it was either as defensive rather than punitive war, on the one hand, or as measures short of war on the other. In the last section, I introduced the idea that some punitive military actions do not involve a large-scale struggle over sovereignty or engage in the widespread targeting of enemy combatants and accept col-lateral harm to non-combatants and are therefore not wars – many of the military actions cited by Kemp fall in this category. The strikes against the Iraqi Mukhabarat in 1993, Gaddafi in 1986, Syria's Shayrat base in 2017, and the Beirut airport by Israel in 1968 were all limited reprisals, measured in their quantity of force and target selection. I discuss the ethical para-meters of reprisals and law enforcement actions below.

There may be cases in which large-scale incapacitation and deterrence appear valuable and even necessary, thus justifying a war. In most of these cases, I argue, such wars are actually defensive rather than merely punitive. Use of force is defensive if it is used against an enemy with whom one is engaged in an ongoing violent conflict, and the use of force seeks to defeat that enemy and stop its ongoing attacks. In my view, the US invasion of Afghanistan after the 9/11 terror attacks was a defensive attempt to stop Al Qaeda attacks from there. Similarly, the fight to overthrow Germany and Japan to disarm them and disband their regimes at the end of World War II could be considered defensive even if the Axis powers' previous aggressive conquests had been overturned. The states involved had been launching attacks, and the push to pacify them is not necessarily punitive, though it may well have rehabilitative, deterrent, and expressive as well as incapacitating effects. Although defensive war can sometimes be unnecessary or disproportionate, because it targets forces currently engaged in aggression it attacks those who are generally liable to attack, in a situation where the benefit of immediate defense makes military force *prima facie* proportionate. Thus, defensive war is generally justified in principle, unlike punitive war.[11]

A complicating factor is that defensive uses of force may seek to achieve ends which overlap with the goals of punishment. This is especially true of incapacitation, which is the primary means of defense. Those who engage in defense frequently hope to achieve deterrence, rehabilitation, and retribution in the course of their defensive struggles as well. If the war is justified by defense, these additional punitive motivations are not problematic unless their pursuit leads to additional violence which lacks a defensive cause.

The US invasion of Mexico to pursue Villa also probably has to be considered a war due to its size and extension to a struggle between the two states. In that case, I concur with Kemp that it lacked defensive justification. Evidence suggests that Villa lacked the capacity and inclination to repeat actions such as his Columbus raid, regardless of the Pershing-led US retaliation (Welsome 2006). Thus, the US action was punitive rather than defensive and, in my view, was unjustified. This is not a counterintuitive result. Historical commentators have criticized the US invasion as excessive and imperialistic (Welsome 2006), implying the validity of my judgment. The action only could be justified if there was an expectation that Villa was bound to continue launching deadly attacks and that Mexico was complicit in harboring him, forfeiting its rights against intervention. In this event, the war to stop Villa, infringing Mexico's sovereignty, would have been principally defensive – like the post 9/11 Afghanistan invasion – rather than punitive.

3.10 Objection and reply

To take up objections, one might wonder if the matters I just discussed, excessive damage and insufficient benefits are irrelevant to the criterion of just cause and only relate to the evaluation of proportionality. Some would reject my second, magnitude and proportionality criterion for just cause and follow Augustine and Aquinas as viewing any "wrong received" as just cause for war. The effects of war, including the costs and benefits of punitive means and aims, would be considered as tipping the balance for or against individual punitive wars.

On the contrary, I maintain that for there to be just cause, there must be a large-scale injustice to be combated, and the action taken must have significant value that could, in principle, justify the harms of war. If magnitudes were not required for just cause, a minor theft could be just cause for war. Lest one think that perhaps only a large magnitude of injustice but not a large degree of benefit is required for just cause, imagine that a war were fought in response to an injustice but had no hope of stopping that injustice or achieving any other benefit. It would be simply retaliation for the sake of vengeance. I take it that this war would lack just cause, demonstrating the necessity of a beneficial purpose. Most theorists have accepted that there is an element of proportionality within just cause. We have seen punitive war generally fails this criterion and cannot in itself be a just cause.

A related objection is that even if punitive warfare is generally unjustified, this doesn't rule out individual cases of punitive wars that might meet the just cause criteria. Conceivably, a particular punitive conflict could be known or at least reasonably expected to have sufficient value and also target the responsible sufficiently narrowly, and be legitimately authorized, such that it could be morally justified. Perhaps we should simply assess individual wars and not make blanket rulings on classes of warfare such as "punitive" (or "defensive" or "preventive").[12]

In response, I would defend thinking of only certain types of war as justified, based upon the general deontological rationale and the rule utilitarian consequences of fighting in those conditions. I argue that we should only treat a situation and war purpose as a justifying cause if fighting in that situation and with that purpose tends to have beneficial results. In recognizing aims and conditions and constituting a just cause, we must take into account the range of wars likely to be fought under that cause. This is an application of the view that just war theory should seek to articulate usefully applicable norms that can be followed with desirable consequences. Since punitive war almost inevitably involves violence disproportionate to its benefits and targeting of those not liable to attack, it should be rejected as a just cause. A principle which states that punitive wars can be fought when beneficial would invite the too frequent and excessive use of force. It is better to say that only defensive wars and narrowly targeted punitive reprisals short of war (as I discuss in the final section) are justified.

Another potential objection is that I have refuted punitive war by defining it away, calling responsive measures either defensive or non-war. If so, I have effectively accepted most of what is typically thought of as punitive while only quibbling about terminology, and my argument lacks normative implications. In reply, I emphasize that there can be punitive wars beyond mere reprisals, which are not defensive. They may be solely retributive, or aimed at general deterrence, or aimed at specifically deterring a party that has no ongoing attack plans. The Iraq War and the invasion of Mexico over Villa may have been punitive wars, and I have argued were to that extent unjustifiable.

Some defenders of punitive war may mostly mean to support what I have called "reprisals," such that our normative differences are less than I have suggested. However, I would point out that the terminology of "punitive war" is apt to be understood to apply to relatively large-scale conflict beyond targeted strikes. It is important to explicitly reject this, which is what I have tried to do decisively here. In the next section, I will qualify my advocacy of reprisals as well, stipulating moral conditions and concluding that most are unjustified.

A final objection is that some of my points about the excesses of punitive war apply to defensive wars as well. War, in general, tends to involve the targeting of parties beyond their liability as rights violators. Additionally, its benefits are uncertain, while violence has certain harm and tends to be excessive. Nonetheless, when an attack is directly defensive, the relative certainty of the attacker's liability and the greater likelihood of furthering the good and avoiding wrongdoing by resisting makes defensive war rightly a different category of action, such that it can be considered *prima facie* a just cause. A norm permitting defense is unlikely to lead to instability and excess violence through spreading attacks, in contrast to punitive war. I hasten to note again that defensive war is not always justified. Although defensive war in principle is proportionate and *prima facie* justified, one must determine whether particular instances of it are proportionate and otherwise justified before approving them.

3.11 Ethics in measures short of war

I have argued that insofar as punitive uses of force can be just causes, these are measures short of war. These strikes can take various forms. Two general paradigmatic forms are that of reprisal and that of law enforcement. Reprisals are undertaken by a harmed party in response to a (perceived) injustice, with the goal of stopping the party from continuing similar acts. They are limited measures, typically proportionate to the wrong to which they respond.[13] The goal is defensive in that the party attempts to prevent more wrongs against itself by specific deterrence and perhaps a degree of incapacitation. However, the reprisal is not directly defensive insofar as it doesn't attempt to stop ongoing attacks but rather responds after the fact. The Israeli raid on Beirut Airport, US strikes on Gaddafi in Libya and Iraq's Mukhabarat, as well as the Clinton Administration's attacks on Pakistani Al Qaeda camps and a Sudanese chemical plant after terror attacks, were all reprisals. The 2020 US strike killing Iranian General Soleimani along with 10 other Iranian and Iraqi officials and militia members was a reprisal for Iranian and Iraqi militant attacks on US bases in Iraq, including a recent one that had killed a civilian contractor (BBC News 2020). The Iranian response to Soleimani's killing, launching missiles at a US base in Iraq, causing injuries but no fatalities, was also a reprisal in turn. Reprisals are not necessarily precisely targeted at those responsible for wrongdoing, hence Walzer's summary of them as "deterrence without retribution" (1977, 207).

Law enforcement measures are taken to punish breaches of international norms. They are paradigmatically not undertaken by the victim and attempt to punish those specifically responsible. They may seek specific or general deterrence and incapacitation as well as retribution for particular wrongdoers. The raid to capture Noriega could be seen as law enforcement. The recent 2017 US strike on the Shayrat airport in Syria for its chemical weapons use (BBC News 2017) was also a sort of law enforcement strike. The same might be said of attacks on the Iraqi nuclear program by Israel and the US in the 1980s and 1990s. One could think of the bin Laden raid as a matter of law enforcement since it seemed to in part seek punishment for the Al Qaeda leader's past crimes. However, since it was arguably was part of an ongoing conflict between Al Qaeda and the US and undertaken by the victim state, it could be considered part of a larger defensive war. In any case, some of the drone strikes and other targeted killings undertaken as part of the "war on terror" could be seen as punitive law enforcement measures.

Both reprisals and law enforcement measures are candidates for justified punitive force because they avoid war's full-scale destructiveness and tend to focus more narrowly on those responsible for crimes and potentially liable to attack. Without these measures, as Leiser (1975, 162) complains, victim states would seem to be forced to "make sitting ducks of themselves and of their people in order to adhere to the letter of the [UN] Charter." He continues that in current international relations, "nations often have no alternative but to resort to measures of self-help." Leiser (1975, 163) concludes and states, including the US and Israel, have also maintained that reprisals are legal under Charter Article 51, guaranteeing the inherent right of self-defense. Law enforcement measures are also clearly legal insofar as they are authorized by the UNSC as Chapter VII measures to uphold international peace and security.

However, reprisals and law enforcement strikes are not unproblematic. They are not directly defensive. Reprisals, inherently unilateral, are almost never internationally authorized. Even law enforcement measures in practice rarely secure UNSC authorization. Most international legal rulings and scholars consider armed reprisals illegal (Cassese 2005). Exceptions have been made for attacks that immediately respond to enemy aggression, e.g. firing back across the border (Cassese 2005). While not directly defending against the attack, by deterring its continuation such measures are the most plausible extensions of defense. Nonetheless, armed force is likely to cause casualties who bear little if any responsibility for the criminal action. Moreover, reprisals and enforcement strikes commit what is typically considered cross-border aggression, thus undermining international norms of self-determination and international peace and security. They can lead to a sequence of tit-for-tat reprisals, as each party considers the enemies' last attack unjust aggression. The reprisal involves engaging in a violation of just war norms on the grounds that someone else has done it first. The difficulties in determining what is proportionate, non-excessive, and effective reprisal leads Walzer (1977, 207) to caution that, "No part of the war convention is so open to abuse, is so openly abused." The US assassination of General Soleimani could have been viewed as disproportionate and risked continuing escalation. The Iranian response risked this as well, though its restraint to a non-lethal, quite clearly proportionate (although it was reported to cause brain injuries to a number of US troops) measure and the resulting US decision not to continue the cycle brought the parties back from the brink of deadly escalation.

Walzer justifies a class of reprisals that strictly uphold non-combatant immunity, on the model of the Israeli Beirut airport raid, which was de-signed to avoid casualties. Attacks on property violate the rights of some innocents and infringe the state's sovereignty rights. Arguably, private owners could be compensated by the state, minimizing injustice. Property rights, unlike the right to life, can be overridden by the value of providing deterrence of terrorist attacks and aggression. The violation of sovereignty and the financial hardship imposed on a poor country is troubling. If a state is merely unable to stop attacks from issuing from within its borders, it should be given support rather than weakened through bombing. Insofar as a state is culpably supporting terrorism or refusing to make a good faith effort to prevent it, this punishment could be justified.

That reprisals will generally be exacted by the strong against the weak and are decided on unilaterally by an aggrieved party judging its own case makes them troubling, even as they are preferable to full-scale war and can more plausibly uphold rights at proportionate costs. In several of the US cases above, it was unclear whether the correct perpetrator was identified and struck. For example, there was controversy about the identification of Iraq in the assassination attempt on Bush and in Clinton's link of the Sudanese chemical plant to terrorism. The lack of certainty combined with the lack of vetting by an impartial adjudicator makes these actions dubious.

Law enforcement strikes are preferable insofar as they are not under-taken by the victim and are done in the name of international norms. Strikes that seek to apprehend or kill criminals responsible for terrorism or which seek to destroy or degrade a state or group's illegal weapons ca-pacity are candidates for justifiable measures short of war. They do not need to meet the threshold of massive rights violations which are required to justify war. However, these strikes tend to pose risks to innocents and can also cause escalation and instability. Although done in the name of law enforcement, they have generally been undertaken unilaterally, or at least without UNSC or other international authorization. This raises problems of bias, ulterior motives, epistemic limitations, and possibly excessive, disproportionate, or indiscriminate force. As a result, such attacks tend to compound rather than resolve injustices.

Indeed, many of the enforcement strikes, as well as reprisals, appear to be dubious. Some have argued that the US was spying on Iraq through its weapons inspectors and acted to provoke the justification for its 1998 attack. The capture of Noriega was a major infringement on Panamanian sovereignty, continuing a history of US interference in the affairs of that and other Latin American countries. There is a concern that such actions have been driven more by US ends than the pursuit of justice.

Law enforcement measures should be guided by international authorization on the basis of an impartial determination of an international crime. Measures that bypass international authorization should be done only if the international authority is unwilling to do due diligence in considering the matter. As law enforcement, these should attempt to scrupulously avoid harm to non-combatants (not even accepting collateral damage on double effect grounds), should be proportionate, and should minimize harm. When possible, criminals should be captured and tried rather than summarily killed, as was done in the bin Laden raid and an unknown number of drone strikes. These stipulations are consistent with just war criteria of discrimination, proportionality, and necessity, albeit made stringently as appropriate for non-war, law enforcement measures.[14]

3.12 Conclusion

Punitive war is unjustified because it targets and harms those who are not liable to attack and because it is very likely to do little good and disproportionate harm. It is at odds with the ideals of punitive justice that seek to justify it as well as the just cause concept. While there are wars that appear to be justified and have a punitive element, we saw that these are only justified insofar as they are defensive. Plausibly justified punitive force can only take the form of reprisal and law enforcement measures short of war. However, these measures themselves can only be justified if they are planned to avoid killing anyone other than those convicted or identified as guilty by legitimate procedures and who cannot be otherwise apprehended.

Proponents of punitive war and reprisals will surely protest that my view leaves states with insufficient recourse in the face of attacks short of war, including international terrorism. However, I would argue that nonlethal measures and narrowly targeted law enforcement are real options and avoid the immorality of unregulated lethal strikes, much less full-scale wars seeking punishment. I would also reemphasize that punitive war is actually unlikely to be effective if limited, such as to be reasonable in cost. The intuitive sense that punitive force must be acceptable is based upon a desire to feel like one is doing "something" in response to injustice, along with the vengeful desire to have someone else suffer in turn. These widespread sentiments are understandable, but they cannot provide ethical justification for killing.

Notes

1 I discuss Locke and Grotius' arguments along with Kaplan's below.
2 I take it that vengeance out of the anger of having suffered harm is clearly not a just cause. Insofar as it lacks a conception of justice in its response, it does not even meet the definition of punitive war and is not supported by most punitive war proponents.
3 For simplicity's sake, I treat rehabilitative and expressive rationales for punishment as part of the utilitarian rationale, whereas some would distinguish these. Separate from utility, I don't think rehabilitation or expression can go far toward justifying the violence involved in war.
4 As I argue below, the deterrent benefits of war are questionable and unlikely to be proportionate to its harms.
5 In Chapter Seven on the ethics of war continuation, I argue against Hurka's view that a war can be continued just for the sake of the contingent causes after the completion of the independent cause. Here I criticize general liability and punitive war, while there, I emphasize the need to meet the threshold of *jus ad bellum* throughout a war.
6 Kemp seeks to justify the Mukhabharat attack, using the others as examples. I take it that now this attack is another candidate for a justifiable punitive strike.
7 Lang (2005) lists additional interventions as punitive, including the US interventions in Haiti (1994), Somalia, and Iraq (1993), as well as the French intervention in Rwanda. In my understanding, these interventions aimed to stop ongoing injustice and not merely punish it after the fact, so they were *defensive* rather than merely punitive. Some uses of military force may be aimed at or motivated by a desire to both defend against wrongful attacks and inflict retributive or deterrent punishment. As a mere additional motive without additional warfare, this doesn't make the war problematically punitive. It is when punishment is the primary justification for war that it becomes problematic.
8 Below I discuss the problem that many of these strikes seem to have unjustly targeted or harmed innocent parties as well.
9 Luban (2011, 313–7) adds that states are traditionally not recognized as having the authority to judge their peers. This objection would not apply as directly to the punishment of non-state actors like terrorist groups.
10 Punitive war might be argued to fall under Chapter Seven prerogative of the Security Council to authorize force to protect against threats to international peace and security. Although there are legal arguments against punitive war, the law is probably sufficiently open to interpretation that the legality of punitive war may largely hinge on the moral understanding of whether it is a just cause.
11 I further defend defensive war as more readily justifiable than humanitarian intervention in Chapter Four.
12 This sort of objection is developed by Allen Buchanan (2007) in his critique of what he calls the "bad practice" argument when types of war are rejected for their generally harmful consequences.
13 For a definition and defense of reprisals, see Leiser (1975). For an account with more stringent restrictions, see Walzer (1977, 207–22).
14 Reprisal and law enforcement force short of war is, in a sense, not a last resort, since full-scale war is the last resort. However, last resort applies insofar as nonviolent measures should be tried before armed force.

References

Augustine. 2006. "Questions on the Heptateuch." In *The Ethics of War: Classic and Contemporary Readings*, edited by Gregory Reichberg, Henrik Syse, and Endre Begby, 82–3. Malden, MA: Blackwell.

BBC News. 2017. "Why was the Shayrat Airbase Bombed?" April 7, 2017. Accessed November 1, 2020. https://www.bbc.com/news/world-us-canada-39531045.

BBC News. 2020. "Qasem Soleimani: U.S. Kills top Iranian General in Bagdhad Airstrike." January 3, 2020. Accessed November 1, 2020. https://www.bbc.com/news/world-middle-east-50979463.

Bedau, Hugo Adam, and Erin Kelly. 2019. "Punishment." In *The Stanford Encyclopedia of Philosophy* (Winter 2019 Edition), edited by Edward N. Zalta, https://plato.stanford.edu/archives/win2019/entries/punishment/.

Buchanan, Allen. 2007. "Justifying Preventive War." In *Preemption: Military Action and Moral Justification*, edited by Henry Shue and David Rodin, 126–42. Oxford: Oxford University Press.

Cassese, Antonio. 2005. *International Law*, 2nd edition. Oxford: Oxford University Press.

Claude, Inis. 1966. "Collective Legitimization as Political Function of the United Nations." *International Organization* 20 (3): 367–79.

Feinberg, Joel. 2014. "The Expressive Function of Punishment." In *The Philosophy of Law*, 9th edition, edited by Joel Feinberg, Jules Coleman, and Christopher Kutz, 789–800. Boston: Wadsworth.

Feinberg, Joel, Jules Coleman, and Christopher Kutz, eds. 2014. *The Philosophy of Law*. Boston: Wadsworth.

Franck, Thomas. 1990. *The Power of Legitimacy among Nations*. Oxford: Oxford University Press.

Hurka, Thomas. 2007. "Liability and Just Cause." *Ethics & International Affairs* 21 (2): 199–218.

Kaplan, Shawn. 2013. "Punitive Warfare." In *Routledge Handbook of Ethics and War*, edited by Fritz Allhoff, Nicholas Evans, and Adam Henschke, 236–49. New York and London: Routledge.

Kemp, Kenneth W. 1996. "Punishment as Just Cause for War." *Public Affairs Quarterly* 10 (4): 335–53.

Lang, Anthony. 2005. "Punitive Intervention: Enforcing Justice or Generating Conflict." In *Just War Theory: A Reappraisal*, edited by Mark Evans, 50–70. New York: Palgrave Macmillan.

Lee, Steven. 2012. *Ethics and War*. Cambridge: Cambridge University Press.

Leiser, Burton. 1975. "The Morality of Reprisals." *Ethics* 85 (2): 159–63.

Leung, Rebecca. 2004. "Clarke's Take on Terror: What Bush's Ex-Advisor Says about Efforts to Stop War on Terror." *CBS News Report*, March 19, 2004. Accessed online October 11, 2020. https://www.cbsnews.com/news/clarkes-take-on-terror/

Luban, David. 2011. "War as Punishment." *Philosophy and Public Affairs* 39 (4): 299–330.

May, Larry. 2008. "Just Cause." In *War: Essays in Political Philosophy*, edited by Larry May, 49–66. Cambridge: Cambridge University Press.

McMahan, Jeff. 2008. "Aggression and Punishment." In *War: Essays in Political Philosophy*, edited by Larry May, 67–84. Cambridge: Cambridge University Press.

Orend, Brian. 2013. *The Morality of War*, 2nd edition. Buffalo, NY: Broadview Press.

Walzer, Michael. 1977. *Just and Unjust Wars*. New York: Basic Books.

Wellman, Christopher H. 2012. "The Rights Forfeiture Theory of Punishment." *Ethics* 122 (2): 371–93.

Welsome, Eileen. 2006. *The General and the Jaguar: Pershsing's Hunt for Pancho Villa: a True Story of Revolution and Revenge*. New York: Little, Brown and Company.

Part II

Defense of a legalist just cause threshold

4 Against the new cosmopolitan interventionism: why human rights protection is not sufficient cause for war

4.1 Introduction

I have argued, against realism and hawkish neo-traditionalism, that there is a presumption against war which can only be overridden if all the *jus ad bellum* conditions are satisfied. However, neo-traditionalism is not the only or even the most influential trend in just war theory arguing for the more ready acceptance of military intervention. The predominant alternative to legalism comes in the form of a cosmopolitan liberalism, which rejects the traditional norm of state sovereignty and conceives of the violation of individual rights as just cause for war. Proponents of this new "human rights paradigm"[1] reject the national defense paradigm, even in the modified Walzerian form, as overly restrictive and morally incoherent. With its intuitively appealing moral universalism and straightforward, consistent conceptual framework, the cosmopolitan human rights view is unquestionably ascendant among just war theorists.[2]

Nonetheless, the justification of war for human rights is, I will argue, a non-sequitur as well as counterproductive. I begin this chapter by describing the emerging human rights paradigm (HRP) and its criticism of the defense paradigm. I proceed to argue against the HRP, rejecting its conceptions of legitimacy, sovereignty, and the right of intervention. I argue for the relevance of magnitude and proportionality of injustice in the conception of just cause, and I defend an asymmetry between just cause for revolution and intervention. I conclude that a plurality of considerations individually and collectively point toward retaining a modified national defense paradigm along the lines articulated by Walzer. I end by defending the defense paradigm against cosmopolitan objections.

4.2 The cosmopolitan critique of the non-intervention norm

From the initial release of *Just and Unjust Wars*, cosmopolitan critics have charged that sovereignty does not have the moral weight that it is accorded in the legalist defense paradigm. Walzer held that national self-determination paradigmatically justifies war, such that defense of a state

DOI: 10.4324/9781003105381-4

or people's sovereignty is a just cause and intervention infringing sovereignty is almost always unjust aggression. Cosmopolitans respond that defending existing flawed states and borders against aggression is neither necessary nor sufficient to justify war. The core of the cosmopolitan argument is that not state sovereignty, but rather the protection of human rights, is the foundation of international justice, including just recourse to war.

State sovereignty, the cosmopolitans convincingly argue, has no intrinsic value. It is only valuable insofar as it promotes the rights and well-being of individuals, a point even Walzer acknowledges.[3] Sovereignty can and, it is generally admitted, frequently does have instrumental value by protecting a sphere of self-determination. Walzer (1977, 54) argued that even repressive, undemocratic states should be assumed to promote a national political culture representing the "common life" of the people, deriving legalist state self-determination rights from individual autonomy rights. However, cosmopolitan critics respond that Walzer's statism assumes a national unity non-existent in culturally and politically diverse modern societies. David Luban (1980b) influentially dismissed the Walzerian presumption of sovereignty as a fallacious "romance of the nation state," while Gerald Doppelt (1980) characterized it as "statism without foundations." Walzer's view that a government will fit the people seems to underestimate the power of authoritarian governments to rule in ways largely unaccountable to their citizens. As Luban eloquently puts it, sometimes "the government fits the people the way the sole of a boot fits a human face" (1980b, 396). Cosmopolitan human rights proponents welcome Walzer's recognition that "the greatest danger that many people face today is from their own states" (2000, xi) but think he fails to infer the logical conclusion of jettisoning the strong presumption of sovereignty. Treating a border crossing as wrongful interference with a sovereign state, explains liberal political theorist Charles Beitz, assumes that national politics further a people's self-determination "simply because it is a domestic... process" (1980, 388). This lacks moral foundation and appears too sanguine a view of the world's states. Even if a state is in some sense an expression of a people's national culture and thus their right of political self-determination, one can still ask, with Doppelt (1980, 403), "Why should Walzer's individual right to national autonomy be more basic than other human rights, such as freedom from terror, torture, material deprivation, illiteracy, and suppressed speech?"

Critics note other seeming incoherencies in the defense paradigm. Lee (2012, 75–6) argues that the framework is plagued by ambiguities about what counts as aggression and, hence, is a sufficient cause for war. It does not seem that every violation of sovereignty, especially minor border crossings or other infringements of self-determination, such as treaty violations or closed shipping lanes, amounts to a cause for war. David Rodin (2002) takes the HRP further, arguing that national defense in itself is not sufficient cause for lethal force; only violations of individual rights suffice.[4]Larry May (2008) concurs that whether aggression should be treated as just cause depends on the manner and magnitude of the threat to human beings posed: "Protecting state territory is not enough to warrant the taking of human life that is an almost inevitable part of war. Invasion of territory does not necessarily mean that innocent people are attacked" (55). For example, the seizure of uninhabited territory – such as China taking over the Daioyu/Senkaku Islands historically controlled by Japan (to invoke a possible case) – is probably not just cause for war.

There is another ambiguity about the defense paradigm's demarcation of aggression and defense, which Lee (2012, 75–6) calls the "time lapse problem." The norm of national defense is thought to include a right to retake territory lost to an aggressor, such as the liberation of Kuwait from Iraqi occupation in the 1991 Gulf War. However, at some point, a victorious aggressor becomes the defender of a new status quo, which the defense paradigm would allow to be protected. For example, the state of Israel and the southwestern US include territory arguably initially acquired by aggression against Palestine and Mexico, respectively, but are now widely accepted as under the sovereign control of the new occupant. Indeed, most of the territory in the world has at some point been seized by force or demarcated by arbitrary processes. The anti-interventionist legalist framework rests upon boundaries but gives little basis for establishing them in a non-arbitrary way. Walzer's (1977, 55) proposal to resolve disputed sovereignty with a popular vote in the territory ("the land follows the will of the people") seems oversimplified and implausibly destabilizing.

The most damning moral objection to the defense paradigm is that it prohibits humanitarian intervention. On a strict legalist view, a state can commit any degree of human rights violations within its own borders, up to and including genocide, without being a candidate for intervention so long as it does not violate the sovereignty of other states. Indeed, some read the UN Charter to strictly prohibit humanitarian intervention (Cassese 2005, 373–4).[5] The cosmopolitan interventionist points out that this is extremely counterintuitive; preventing genocide or other massive violation of human rights would seem to override the value of sovereignty. Most commentators agree that India was justified in intervening to stop the assault on human rights in East Pakistan (Walzer 1977, 105–6) and that the world community failed morally in not acting more robustly to stop the genocide in Rwanda. The near moral consensus about this appears to refute the defense paradigm.

Walzer (1977, 101–8) permits humanitarian intervention in extreme cases that "shock the conscience of humanity," such as "genocide or mass enslavement." However, critics point out that this exception fits uneasily with his general basis of war's justice on state self-determination. Cosmopolitans wonder why national self-determination, if basic, does not permit even genocidal states. On the other hand, if those human rights abuses can outweigh or negate sovereignty, critics wonder why rights violations short of those atrocities could not do so as well. Lee (2012, 117) concludes that there is incoherence in the idea of humanitarian intervention as an exception to the general rule of national defense. The exception implies that it is really human rights protection and not national defense that is the basis of justice in war.

With the recent UN approval of the Responsibility to Protect (R2P) norm, international law is moving in the direction of the HRP. The R2P explicitly allows for humanitarian intervention in cases that do not necessarily pose threats to foreign states. At the 2005 World Summit, the General Assembly agreed unanimously on new language specifying four atrocity crimes as justifying intervention: "Genocide, war crimes, ethnic cleansing, and crimes against humanity" (UN General Assembly 2005, par. 139).[6] Insofar as the R2P norm is an agreement about the application of conditions for intervention under Chapter VII of the UN Charter and has begun to be referenced in UNSC resolutions and statements by the Secretary-General and other authorities, it could be argued to be new international law.[7] As I discuss in the next section, proponents of the HRP tend to think that the R2P is not permissive enough in its limitation of intervention to mass atrocities.

A final objection to the defense paradigm is that it does not provide moral guidance for the just recourse to civil war. Civil wars are now the most common conflicts and are responsible for more deaths in recent decades than interstate wars. The legalist prohibition of aggression between states says nothing about armed struggle over governance within the state. International law does not prohibit internal uses of force or give guidelines for its justification. Walzer's legalism accepts civil wars as part of the process of state self-determination. His only limiting rule is that foreign intervention is prohibited unless it counter-balances previous intervention on the opposite side (Walzer 1977, 96–101). This oddly amoral view is problematic in accepting deadly wars as ordinary, in recognizing equal rights to fight regardless of cause and purpose, and in portraying the outcome of force as legitimate self-determination. It is not clear why outside parties should not try to help the more rights-respecting party. Restricting intervention to Walzerian "counterbalancing" assistance may perversely prolong and escalate conflict without resolving it (Lee 2012, 246–54).

4.3 The case for the human rights paradigm

Cosmopolitan interventionists argue that protecting individual human rights is necessary and sufficient for just cause for war. Human rights have been increasingly recognized globally, such that they form the basis of an understood moral minimum that ought to be provided for each individual. The 1948 Universal Declaration of Human Rights and treaties establishing particular rights, such as the Geneva Conventions and Genocide Convention, give human rights institutional recognition. "Human rights" are regularly invoked by both states and non-governmental organizations. Although some object that human rights are unique to secular or Western culture, there is a strong case to be made that there is an overlapping consensus on belief in human dignity and a need for autonomy rights as a means to any particular valued ends (Runzo, Martin, and Sharma 2003). As I argued in the Introduction, moral argument and practice concur on the importance of human rights.

Cosmopolitans argue that state legitimacy, the right to exercise political power, is conditional upon minimal justice in terms of upholding basic rights. According to the definition provided by Allen Buchanan, probably the most systematic and influential cosmopolitan theorist of legitimacy:

> A wielder of political power .... is legitimate (morally justified in exercising ower) if and only if and only if it (1) does a credible job of protecting at least the most basic human rights of all those over whom it wields power and (2) provides this protection through processes, policies, and actions that themselves respect the most basic human rights (2007, 247).

Thus, states which fail to systematically uphold basic human rights are illegitimate and unjustified in exercising power.

On the cosmopolitan understanding, human rights are valid claims that each person has against the world community. Although the primary responsibility for meeting these rights falls to the local state and government, if they fail to protect and respect rights, the international community – like the federal government in a system of states – should step in to protect those rights. This is explicitly legislated in the R2P, at least for atrocity crimes:

> The responsibility to protect acknowledges that the primary responsibility [for protecting human rights] rests with the state concerned and it is only if the state is unable or unwilling to fulfill this responsibility, or is itself the perpetrator, that it becomes the responsibility of the international community to act in its place. (International Commission on Intervention and State Sovereignty 2001, 17)

The cosmopolitan hawk makes the additional claim that illegitimate governments and their states lose their right of sovereignty, including their right against foreign invasion, such that just cause for war presumptively exists against such groups. In his classic defense of humanitarian intervention, Teson (1988) unambiguously states this connection between governmental injustice, illegitimacy, loss of sovereignty, and just cause for intervention:

> My main argument is that because the ultimate justification of the existence of states is the protection and enforcement of the natural rights of the citizens, a government that engages in substantial violations of human rights betrays the very purpose for which it exists and so

forfeits not only its domestic legitimacy but its international legitimacy as well. Consequently ... foreign armies are morally entitled to help victims of oppression in overthrowing dictators, provided that the intervention is proportionate to the evil which it is designed to suppress.... [H]umanitarian intervention is justified not only to remedy egregious cases of human rights violations, such as genocide, enslavement or mass murder, but also to put an end to situations of serious, disrespectful, yet not genocidal, oppression (15).

Because the government has lost legitimacy and other members of the global community have a natural duty of justice to uphold human rights, these others have an obligation – and thus a right – to use force to protect human rights in the illiberal state.

In his more recent work, Teson drops the suggestion that it is the loss of sovereignty *per se* that permits intervention. Rather, state borders are merely irrelevant such that "sovereignty and self-determination play no role in the justification of humanitarian intervention" (2017, 48). Injustices should be stopped, and if the home state refuses to stop them, others have a right to do so, and the home state has no right to prevent them. Teson defends this idea of the irrelevance of sovereignty with a thought experiment in which the push of a magical green button would overturn foreign human rights violations without harming anyone. His intuition is that there would be no reason not to push this button to undo the injustices around the world. "Costless intervention is *always* permissible against a government that violates the rights of its subjects" (46). Teson suggests that even interfering with a generally legitimate government is justified if proportionate: if we could free the millions of unjustly incarcerated people in the United States with a push of the button, "we may – indeed must – press the button" (47). From the intuitively unproblematic justice of costless intervention, Teson infers that "[t]he principle of state sovereignty ... has nothing to do with the impermissibility of intervention" (2017, 45–6). He also notes that diplomatic pressure to end human rights abuses is generally accepted, inferring that infringement of state self-determination is not inherently problematic.

In addition to the case for intervention based upon international responsibility, there is an argument from the right of defense. It is generally accepted that oppressed people have the right to liberate themselves by force. Moreover, it is commonly accepted that one might voluntarily transfer one's rights to others to exercise them on one's behalf. This implies that if a group is denied their rights to the extent that rebellion would be justified, and if they welcome intervention to end their oppression, foreign intervention would be justified. As Buchanan puts it, "if war is morally permissible for the sake of establishing democracy for ourselves, could not war to establish democracy in another country that is so thoroughly repressive as to make revolution virtually impossible also be a moral option?" (2010)[8] Ned Dobos' (2012) *Insurrection and*

Intervention systematically interrogates the traditional legalist "asymmetry thesis," that intervention is less readily justified than revolution, and argues that it is largely unsupported despite its prevalence. He concludes that

> If we are prepared to concede that a people would be justified in rising up violently against their government, then unless we have reason to believe that foreign intervention on their behalf would be a predictable failure, excessively costly, or internally illegitimate, we have no good reason to oppose it (208).

The HRP seeks to correct this seeming legalist contradiction by permitting the same deprivation of rights to justify foreign intervention as domestic insurrection.

Recent formulations of just cause show wide support for the HRP. There is internal variation about matters such as whether human rights violations are the only just cause and how serious the human rights violations must be. Luban writes that a just war is "(i) a war in defense of socially basic human rights (subject to proportionality) or (ii) a war of self-defense against an unjust war" (Lee 2012, 128). Orend (2013) and Fotion (2007) concur that national defense is a cause for war alongside the defense of human rights. On the other hand, thinkers including Lee (2012), May (2008), Fabre (2012) and Lango (2014) would eliminate the second clause, arguing that defense of human rights is necessary as well as sufficient to justify war. For revisionists following McMahan (2005; 2009), just war theory should be derived from the same principles as individual ethics, and a just cause for war must be understood in terms of individual liability to attack based upon threats posed to other individuals. This consistent, fundamentalist variant of the HRP is becoming the predominant view among liberal just war theorists.

The largest debate among cosmopolitans is over the magnitude of the threshold of human rights violation that constitutes just cause. In general cosmopolitan thinkers believe the lines set by Walzerian just war theory and international law, even the recent R2P norm, are too restrictive of humanitarian intervention. Instead of a threshold of "crimes that shock the conscience of humanity" or "mass atrocity," cosmopolitans argue that any violation, at least any systematic violation, of human rights is a just cause for war. Some cosmopolitans would limit just cause more narrowly to violations of *basic* human rights. Luban, Lee, and Shue understand basic rights to be those which provide a condition for autonomous agency, including freedom from slavery, the right to life, and the right to the material conditions for subsistence. They would exclude democratic political participation, strict social equality, and a wider range of individual freedoms from the most basic rights.

Other cosmopolitan theorists hold that a broader array of rights violations justify armed intervention. Some define state legitimacy in terms of maintaining a minimally just society, permitting intervention into any society deemed generally unjust. For Orend, "all talk of justice regarding war must revolve ultimately around legitimate governance" (2013, 37). Legitimate governance is defined in terms of being "minimally just," including "mak[ing] every reasonable effort at satisfying the human rights of its own citizens."(38). Orend concludes that "regimes which fail the conditions of minimal justice are not legitimate and have no state rights, including the right not to be attacked and overthrown." Although he indicates sympathy with the Walzerian view that humanitarian intervention is an exceptional justification to be reserved to mass atrocities, his view of the centrality of legitimacy to just cause ultimately contradicts that common sense judgment. Applying his criterion, he ends up concluding that the Hussein regime in Iraq had "no right not be overthrown," though he condemns the actual intervention for not having humanitarian rescue of Iraq as its primary guiding intention (101–3). The idea that the Iraq War had an in principle just cause but was only problematic because of its specific intentions, methods, and practical consequences, is common among cosmopolitan writers.

Kimberly Hudson (2009) argues for what she calls an "Atrocity Standard" following the R2P. However, she argues that since among the enumerated "Crimes against Humanity" is "persecution," and the widespread denial of any right could be said to be a form of persecution, societies that practice discrimination on gender, ethnic, and religious bases would provide just cause for intervention. Giving examples, she suggests that Mugabe's Zimbabwe, along with Afghanistan under the Taliban, would meet her criterion (55–66). Even a Rawlsian decent hierarchical society, because it practices systematic discrimination though also upholding respect for basic rights to life and liberty, "could be so oppressive that there is a just cause for [humanitarian intervention]" (Hudson 2009, 67–8).[9]

For his part, Teson argues that ordinary tyranny, with repression and human rights abuses (alongside anarchy, where human rights are also unprotected) is a just cause for war, and he defended the Iraq War on these grounds (2005). Teson now cautions that in some cases, intervention to stop less than genocidal tyranny will not be justified because it may fail to achieve humanitarian effects, as happened in Iraq (Teson and Van der Vossen 2017). Proportionality and likely success are thus the only bars to intervention to enforce human rights norms.[10]

Against the longstanding view that the political improvement of a foreign society is not cause for war, cosmopolitan theorists have begun to defend democratization as a just cause. It is now common to argue that democratic political participation is a basic human right, that undemocratic states thus fail to be legitimate, and that there is just cause to forcibly remove their governments, replacing them with democracies. As Buchanan puts the case:

a theory of legitimacy that does not include a democratic requirement faces an unanswerable objection: If the political system should express a fundamental commitment to equal consideration of persons, why shouldn't this commitment be reflected in the processes by which laws are made and in the selection of persons to adjudicate and enforce the laws, not simply the content of the laws? (2007, 251).

Democratic theorist Robert Talisse (2012) seconds Buchanan, citing democracy's historic value in providing "systematic and reliable protection to its citizens from war, famine, and human rights abuses." He concludes that democratic governance is either itself a human right or a direct means to human rights, such that its violation serves as just cause for war: "If wars waged for the sake of preventing or stopping human rights abuses and humanitarian disasters are wars with just causes, then wars waged for democratization have just causes, for democratization provides systematic protection from these and other evils" (2012, 197). While Talisse agrees with skeptics that interventions to promote democracy – like the 2003 invasion of Iraq – are typically unjustified, he says this is because of proportionality and last resort considerations and not the lack of an in-principle just cause.

In addition to the *quality* of rights violations required for just cause – there is a question of how many such violations there must be. Most cosmopolitan theorists qualify that there must be a large (using terms such as "systematic" or "widespread") number of such violations to justify war. However, the logic of the cosmopolitan view suggests that even *one* violation of human rights, at least of the most basic kind, can justify war. Lango (2014, 121) explicitly makes this extension in his cosmopolitan just war theory, stating that "there is a just cause for a targeted military operation to stop the killing of a single human being." McMahan similarly asserts that "just cause says nothing about considerations of scale or magnitude, but functions entirely as a restriction on the type of aim or end that may legitimately be pursued by means of war" (2005, 4).

In his defense of the HRP, Lee (2012) argues that human rights do not simply justify some wars in addition to those of national defense. Rather human rights justify all wars, including defensive wars along with intervention. Teson (1988) puts it simply: "war is just if, and only if, it is in defense of human rights." Lee characterizes the logic of this shift in just war views by making an analogy between the relationship of the national defense and human rights paradigm of just cause to that between Newtonian and relativistic physics. While defense, like Newtonian physics, works well for a class of typical cases, it does not work universally. By contrast, human rights, like relativistic physics, explain the morality of all cases, including the standard cases where national defense (like Newtonian physics) works. Thus, it is not that the national defense theory is the general truth and the human rights view an exception. Rather the HRP's view of just cause should be recognized as the ultimate truth (2012, 122).

4.4 Against the human rights paradigm for state legitimacy

Despite its simple and appealing basis in human rights, the HRP must be rejected. Although human rights are worthy ideals of justice, and their protection is central to an adequate just war theory,[11] the straightforward equation of human rights violations with a just cause for war is questionable at several steps and completely implausible in its conclusion. In this section, I will critically assess the reasoning of the cosmopolitan hawks that a government's legitimacy is contingent on its protection of human rights. In the following section, I challenge whether government illegitimacy is sufficient grounds for the loss of state sovereignty and question whether all rights violations are cause for war.

Although the protection of human rights, along with well-being, is the moral basis of government, and there are unquestionably some rights violations that cause governments to lose their legitimacy, it is a mistake to say that legitimacy is forfeit by any and all rights violations. Taken literally, this has the absurd implication that no governments are legitimate since all sometimes violate and fail to protect human rights. Cosmopolitans have a better case when they argue that governments must "substantially," "generally," or "credibly" protect the most basic human rights. Arguably, this allows us to recognize many states as legitimate insofar as they protect a range of liberal rights, do so with at least formal equality of respect for citizens, and according to a publicly established and more or less democratically revisable rule of law.

Nonetheless, there will be ambiguity about what rights are basic and what constitutes systematic failure.[12] In the United States and many other states recognized as liberal democracies, there is great inequality with disparate economic opportunity, life chances, and political power between social classes. There is systematic racism that results in effective inequality before the law and economic and other social discrimination. Laws and institutions frequently discriminate against gays, lesbians, transgender individuals, and other minority groups. A case might be made that human rights protection in the US is limited and their violations substantial. Although some might be happy with the idea that purportedly liberal states like the US are illegitimate and thereby just targets for overthrow, along with all the clearly illiberal states, I take it a theory that justifies so much violent conflict and provides so little stability is implausible.

It appears more reasonable to condition legitimacy more minimally on the protection of a narrower group of "basic rights," as we saw that moderate HRP proponents endorse. However, this is problematic as well. There is disagreement about just which rights count as "basic." While the right to life is obviously included, all cosmopolitans seek to draw a wider circle of moral minimum. This may include equal protection under the law, positive subsistence and welfare rights, substantial personal liberty, and democratic voting and political participation rights. However, any of these could be argued not to be absolute, and many would allow culture and context to play a role in the specification of what sort of equality and freedom are required. There may be justified exceptions to equality of treatment under the law and democratic voting rights. Democratic voting in elections may not be required for political freedom. Forms of participation along the lines of the consultation of the people present in Rawls' (1999) decent hierarchical society of "Kazanistan" could be legitimate. Although an idealization, aspects of such a traditional model of representation of people by accepted leaders who are committed to the good of the people is a real phenomenon. Some traditional cultural practices such as bridal abduction and female genital cutting (or mutilation) cause objective harm to women but nonetheless have substantial acceptance among those who are subjected to them.

Even if we limited the rights in question to the protection of the right to life, this is violated at times in liberal states, when there is excessive police force, use and mis-implementation of the death penalty, and ineffective effort to protect the lives of groups of citizens. If we give any kind of demanding standard of human rights, this will end up ruling states as illegitimate too widely and too frequently. On the other hand, if we remain reasonably minimalist and say something like states are legitimate unless they engage in large scale violations of the right to life or other infringements of autonomy, akin to slavery, then we come close to the restrictive Walzerian paradigm wherein only genocide and other shocking crimes result in the loss of legitimacy, potentially triggering intervention.

Although the HRP invites us to think of any infringement of human rights as a legitimacy-forfeiting violation, it is important to note that rights infringements are sometimes justifiable. Infringements of freedoms may themselves express the will and values of the people. Norms that deny equal rights to women, repress minority religions, or flout property rights with social redistribution, may be supported by even those whose rights are infringed. Arguably, people have a right to waive their rights voluntarily. Moreover, things that are considered human rights may at times be outweighed by the rights and values of the majority. It is widely recognized that the right of free speech, for example, is not absolute and can be infringed for public safety. Laws in Islamic states prohibiting proselytizing by non-Muslims and laws in Quebec restricting the public use and teaching of English may be important enough for preserving traditional values to trump the rights of freedom of religious practice and speech. If a rights infringement is justified either by voluntary waiver or by pursuit of a greater good, then the infringement is not a violation. If foreign parties are allowed to determine states to be illegitimate because of perceived rights violations, this dogmatically and paternalistically overrides local adjudication of competing values.

Of course, rights infringements are frequently unjustified, neither consistently accepted nor plausibly warranted by their necessity to achieve an overarching value. However, even clear rights violations do not automatically make the government illegitimate. Although these rights-violating laws and practices are unjust and should be changed, the government as a whole may retain legitimacy. A government that is a frequent rights violator or systematically fails to protect some human rights may, on balance, protect the rights and welfare of its population. If it does, it is justified – perhaps even obligated – in continuing to exercise power and thus is legitimate. Of course, it should also stop committing injustices and can be rightly criticized for failing to do so. However, it need not stop exercising power and certainly should not be subject to armed foreign intervention and attempted overthrow, which would undermine its effectiveness, force it to spend lives and resources in defense, and probably exacerbate its rights violations.

A rights-violating government may even be supported by the people whose rights are violated. While they may oppose the norms violating their rights, they may also have good reasons for accepting the government's continued rule. They may reasonably prefer a gradual change from within the framework of that government. Given the value of political freedom and freedom of association, it may best uphold the rights of most locals to leave a rights-abusing state in power. Although natural law thinkers have long held (with some controversy), with Martin Luther King, that "an unjust law is no law at all," they do not say that the government which establishes an unjust law is not a legitimate government. The civil rights movement generally accepted the legitimacy of the US and state governments while protesting segregationist laws.

Some governments' rights-violations may lose the support of oppressed minorities but continue to retain the support of the vast majority of the citizens. Although Walzer's idea of a national common life may be an exaggeration, it is right to say that there is frequently an identification with the government and society, such that institutions preserve the good life as it is culturally understood. Despite valid criticisms, rights-abusing governments such as those in Russia, China, Iran, North Korea, and Saudi Arabia have considerable popular support. Their overthrow, especially by foreign intervention, would be widely opposed. The rights violators' continued governance is then supported by the people's freedom of association as well as utilitarian values maximization. Even a bloodless overthrow of the above governments would be a significant violation of the rights of many people in those societies and the government exercising power on their behalf (Walzer 1980).

Moreover, even governments that do not particularly represent a people provide material stability and services which contribute to the people's welfare and their ability to live and enjoy a degree of autonomy. For example, the state may provide policing, peacekeeping, disaster relief, health care, famine relief, and defense, such that the government, on balance, serves the people, and they would be worse off without it. While its rights violations are unjust and should stop, the government does not lose all rights to control the territory. Importantly, there may not be an alternative body prepared to replace it and govern in a way that better protects the citizens' rights.

If one thinks that the slave-holding US's rights were violated by the British invasion in 1812 and that the kingdom of Kuwait had legitimate authority when Saddam Hussein's Iraq invaded in 1991, one recognizes that systematic violation of particular human rights does not make a state lose its legitimate right to govern, much less be a justified target for intervention. As Chris Naticchia (2005) has argued, the human rights standard of legitimacy would render almost all of the Middle East and Africa, as well as much of Asia illegitimate, targets for intervention by Western liberal democracies.

While the Hussein government in Iraq, the Gadhafi government in Libya, and the Serbian government in Kosovo all were regular violators of political and civil rights, they also provided stability and security, which were undermined by intervention. It is not just Rawlsian decent hierarchical states that should be recognized as legitimate – but any state whose continued governance is preferable to available alternatives. What Rawls (1999) terms "benevolent absolutisms" typically preserve rights and well-being despite being illiberal. In practice, states are not purely liberal, non-hierarchical, benevolent, or absolutist, outlaw or non-outlaw, but some mixture. The fact that they provide some goods, are not purely violators, and the denial of their sovereignty introduces new problems are all reasons to give the benefit of the doubt on legitimacy to almost all of the world's governments.

To give a recent notable example, the 2008 Arab Spring revolutions against authoritarian governments across the Middle East, celebrated by cosmopolitan liberals, have had an overall negative impact on individual rights in the region, as the instability undermined protections and enabled the rise of populist Islamist movements which rolled back the human rights gains made under the previous authoritarian regimes (Esfandiari and Heideman 2015). While the regime of Hosni Mubarak was unquestionably repressive, its ouster has led to successive governments, first under President Morsi of the Muslim Brotherhood and now by the military junta headed by el-Sisi, which have been even worse than Mubarak's.[13] To say that the best available governing body is illegitimate because it falls short of standards that cannot be practically implemented is self-defeating.

Cosmopolitans hasten to add that although illegitimate states do not have a right to rule, it will not necessarily be justified to go to war against them. For cosmopolitans such as Teson, loss of legitimacy means that the justice of intervention is largely dependent on proportionality. Yet, making the sovereignty of so many states contingent upon the practicality of intervention would undermine global stability. As I argue in the next chapter, treating illiberal states as illegitimate and potential targets for intervention undermines global cooperation on the challenges facing the world. Whether or not an intervention is undertaken, the denial of self-determination rights exemplifies and exacerbates colonialism and a mutually destructive clash of civilizations.

Rather than adopting a rigid idealistic cosmopolitan human rights model, I suggest that governments in power should be presumed to be legitimate unless (a) there is a massive expression of public will against that government or (b) there are massive rights violations such that the case for intervention is overwhelming and it can be presumed that the government lacks support. In the event that a mass movement explicitly opposes a government, this overrides the presumption of legitimacy. Such a situation indicates not only a lack of popular support for the government but also the presence of a political movement ready to assume power in a manner that better expresses the will of the people. A state which is engaged in massive violations of basic rights, especially the right to life or basic freedom from enslavement, can also be said to lose its legitimacy, even if there is no apparent resistance. Then, the government's control can no longer be said to further freedom and well-being. However, these cases of loss of legitimacy are exceptional, not commonplace, as implied by cosmopolitan thinkers.

Some will worry that the view of legitimacy I put forward here is too conservative, ceding too much to Hobbesian *realpolitik*. I would reply that to acknowledge most governments as legitimate is not to accept all that they do as justified. They can and should be criticized. In some cases, it would be good to continue to criticize in a manner that may contribute to opposition which in turn eventually nullifies the government's authority. Injustice is apt to generate illegitimacy. However, until that does happen, its right to rule – with the corresponding need to make demands upon and negotiate with that body – continues.

4.5 The presumption of state sovereignty and non-intervention

Its view about the illegitimacy of rights-violating states is only one fallacy of cosmopolitan interventionist just war theory. A second is the idea that once a government is illegitimate due to rights violations, the state or people loses its sovereignty, including its right against intervention. Interventionists are fond of distinguishing the state and people as a whole from the government. However, they commonly fail to extend the implications of this for the justification of intervention. The fact that a state's government is illegitimate and has no right to govern does not mean that foreign intervention in this territory is warranted. Although self-determination implies that an illegitimate government should not be imposed, it also implies a presumptive right against foreign intervention. Most interventionists recognize that in order to be justified, intervention must be welcomed by those on whose behalf it is undertaken (Buchanan 2010, 268; McMahan 2010). Actions that radically change people's life chances and put them at significant risk, as with conducting a medical procedure on a patient, should not be undertaken without the subject's consent, even if thought to be in their interest.[14]

In practice, interventions typically fail this requirement of local author-ization. Interveners rarely actually conduct an in-depth discussion with the people who are to be liberated, undoubtedly in part because of practical limits imposed by the authoritarian regime and communication barriers. US interventions, from the early twentieth century colonial missions to Cuba and the Philippines, to late twentieth century involvement in Southeast Asia and Latin America, to early twenty-first century missions in the Middle East and Africa, have had little connection to the people to be helped and have for this reason failed to effectively promote their welfare, rights, or effec-tively establish the security that were the stated goals. For example, the decision to invade Iraq was heavily influenced by the manipulation of the single Iraqi dissident, Ahmed Chalabi (Ricks 2007, 56–7). Too frequently, interventionists are willing to assume that the people's will accords with them (Teson 2005, 15–6), such that the evidence for the voluntariness of the intervention is nothing beyond the theorist's or intervener's view that it is otherwise justified. The local autonomy requirement becomes empty in practice. Orend (2013, 99) and Frowe (2014) defend intervention without consent with an analogy that in domestic society, police do not need an invitation to intervene to stop a gang that is brutally attacking a person. However, this analogy only works for atrocities in which people are being slaughtered, such that we can infer with near certainty what the victims would want, and there is no time for further consultation. Typical oppres-sion and ongoing rights violations short of an atrocity, which are the cases in dispute between my view and the HRP, are more akin to an abusive re-lationship with intermittent rights violations of varying degrees.[15] In such cases, intervention may not be wanted, may make things worse and is not so pressing that it needs to be done without asking for the consent of the victim. The assumption that intervention will be welcomed ignores that many would rather keep a flawed government than have a war or if there is already civil conflict, an escalation of this war with external force and for-eign invasion and occupation.

A particularly troubling aspect of modern war is that it kills many more non-combatants than combatants (Lee 2012, 165–6). Among the innocents that will be killed in a humanitarian intervention are some of the very people the intervention was designed to save. This means that even if the perpe-trators of human rights violations are liable to attack, the use of war as an instrument of rescue is problematic. Teson uses the doctrine of double effect to argue that as long as collateral damage to innocent casualties is an un-intended, proportionate side effect of an intervention to further human rights, it is merely a rights infringement but not a violation (1988, 98–102). However, morality does not readily accept causing the deaths of innocent individuals for the greater good. The double effect principle has been questioned even for *jus in bello*, as it is puzzling why it is fine to accept the foreseeable effect of killing the innocent when it is absolutely wrong to intend. From this human rights standpoint, any foreseeable, avoidable killing of the innocent should be avoided (Nathanson 2010). Also, the double effect doctrine was traditionally used to justify targeting with a risk to innocents in the midst of a war already begun for a greater cause, but not as justification for entering war. To accept military action which kills the innocent whenever it is expected to preserve rights on balance allows recourse to lethal force too readily as a tool of politics. Typical intuitions suggest that if justifiable at all, innocents can only be sacrificed in order to save multiple times more lives than those spent. In the next chapter, I argue that one generally cannot expect such good results from humanitarian intervention. Here my point is that the presumption against killing the innocent in particular poses a bar to even well-intended interventions.

While cosmopolitan interventionists are right that non-interference cannot be automatically equated with self-determination, there is something to Mill and Walzer's (1977, 86–9) point that freedom must be self-won. The point is partly conceptual – freedom as self-government – and partly empirical regarding the improbability of a temporary violent intervention from the outside making lasting improvements in human rights. While it is conceivable that foreign intervention might allow and encourage internal democratic forces, invasion and occupation by foreign armed forces tend to involve significant foreign control over the state's reconstruction, such that armed intervention is in inherent tension with democratic self-determination (Bohman 2008).

Like the idea that human-rights violating states lose legitimacy, the idea that states with illegitimate governments can be justifiably intervened in has implausible implications. Making sovereignty so extremely contingent is at odds with the international framework of the UN, which is "based on the principle of the sovereign equality of its members" (United Nations 1945, Art. 2.1) and would destabilize the international system. While illegitimate governments should not continue to rule, to say that all such states are legitimate targets for foreign intervention contravenes effective global norms.

It is true that sovereignty is, in a sense, conditional. Absolute, unconditional sovereignty would self-contradictorily imply that sovereign states could invade each other at will (Shue 2003). Norms of national sovereignty are legally and ethically understood to be conditioned on the reciprocal rights and duties of states as members of the international community. Since the presumptive norm of non-intervention is widely recognized and generally furthers welfare and rights, there is reason not to condition it on states' strict upholding of human rights or even possession of a legitimate government. On the contrary, given that intervention tends to violate rights and contravenes international norms regarding self-determination, deontological and consequentialist ethics both suggest that sovereignty is only forfeit and intervention justified to stop large-scale violations of human rights in which the people clearly wish for intervention, and intervention is apt to succeed in protecting human rights. My argument that sovereignty, though not an absolute, has continuing relevance and should not be readily forfeited on cosmopolitan grounds concurs with several political and legal theorists (Cohen 2008; Fassbender 2003; Loughlin 2016). Interestingly, although it is couched in cosmopolitan terms about conditional sovereignty, the R2P's atrocity crime standard for intervention (Bellamy and Luck 2018) is closer to Walzer's exceptional "crimes that shock the conscience of humanity" than to the proposals of most cosmopolitan theorists.[16] Intertwining conceptual and practical considerations point to a modified defense paradigm with exceptional mass atrocity standard for intervention, rather than the reduced cosmopolitan thresholds for force.

What, then, of Teson's use of the green button analogy to suggest that costless intervention is unproblematic and that infringement on sovereignty lends no moral weight to non-intervention? My intuitions about and analysis of this case are completely different from his. Coercing people on a mass scale to comply with one's will as a means to desirable outcomes is a problematic infringement of autonomy. This judgment probably in part stems from a worry that intervening parties might make mistakes or abuse their power and pursue dubious ends with bad results. In his account, the humanitarian intervener is self-authorizing to remake the world without any process of representation or accountability. Teson's benevolent magical manipulator, presented as ethically pure and thus justifying humanitarian intervention by analogy, eerily resembles the paternalistic hyper-control depicted in Ursula Le Guin's dystopian thought experiment, *The Lathe of Heaven* (1971). While Teson presents the interference as one of simply stopping rights violations – preventing murder, rape, torture, repression of speech and association, and unjust imprisonment – in practice, to stop rights violations also means making the potential perpetrators do other things, whether go to jail, get new jobs, rewrite their constitutions, revise school curricula, etc. Decisions about such things presumptively should be governed by local self-determination so that coercively undoing rights violations entails an infringement of autonomy. The infringement of autonomy can sometimes be justified but, contrary to Teson's thesis, is not morally irrelevant.

Teson's analogy of the magic button and armed intervention to common and intuitively acceptable diplomatic pressure is also poor. Diplomatic pressure can range from non- or mildly coercive moral persuasion and shaming criticism to moderately coercive political and economic incentives. Such measures encourage states to make non-rights-violating decisions but not do not completely override autonomy in the manner of either armed intervention or Teson's magical control. At the same time, the green button fails as an analogy to armed intervention insofar as the former lacks the threat of violence involved in the latter. Even if an armed intervention could succeed bloodlessly, there is a presumption against achieving political ends via lethal threats. To capture this element of intervention, we could imagine that Teson or a well-meaning world power acquires an overwhelming new weapon, the threat of which it uses to force the rest of the world to comply with human rights norms. I take it that such a threat is morally dubious (and again invokes dystopian villainy) even if it achieves desirable ends bloodlessly. Our intuitive condemnation of beneficial threats results only partly from the worry that they might turn out badly or be misused (as well as causing psychological harm from the experience of the threat and fear of its abuse in the future). I take it that there is also an underlying sense that violent coercion is at odds with human dignity and presumptively wrong. As Jennifer Welsh (2014, 209) notes, humanitarian intervention's value in saving lives must be weighed against not only the value of self-determination but also the general presumption against violence. Teson overlooks the latter presumption at the same time that he discounts self-determination. It may sometimes be justifiable to use threats and force for humanitarian ends around the world, but it is wrong to say that doing so infringes no rights or values. A robust standard of just cause is required to override the norms of non-violence and self-determination.

An HRP proponent might still object that the general duty to combat injustice, when combined with the lack of internal legitimacy on the part of the target state, creates a *prima facie* right (or even duty) of intervention. However, this is dubious for three interrelated reasons. First, the duty not to harm generally outweighs the duty of beneficence. As we discussed in Chapter 2 in the argument for a presumption against war, killing isn't readily justified by the prospect of causing marginally more good than harm. Government illegitimacy is not sufficiently grave to warrant armed intervention. Secondly, the victims of the current injustice do not necessarily welcome assistance, which, as we saw, refutes many interventions. Finally, the act of intervention generally has negative consequences, foreseeable and unforeseeable, which make it likely to do as much or more harm than good. The harm will include killing innocent bystanders and interfering with the right to self-determination on a large scale. I will elaborate on the typical problematic consequences of intervention in the next chapter. For now, I point out that they provide support for the assumption of sovereignty and its corollary of non-intervention.

4.6 The role of magnitude and proportionality in the just cause threshold

I have argued that state illegitimacy and minor rights violations do not amount to a just cause for armed intervention. However, more argument is required to justify the mass atrocity standard. Cosmopolitans argue that any 'systematic" or "basic" rights violations, or the threat thereof, could serve as just cause for armed intervention. Why must there be injustice at the level of massacre or enslavement to justify intervention?

My general answer is that because war involves large-scale killing, destruction, and destabilization, it requires a large-scale cause. Only the defense of life or the most basic liberty, along the lines of freedom from enslavement or perhaps rape, can justify lethal force. While the threat to one individual can trigger a right to self-defense, war, which involves the large-scale killing of anyone fighting for the enemy group as well as many innocent bystanders, needs a much higher threshold of just cause. An extreme means requires a proportionately extreme cause. This is founded on both deontological principle and utilitarian cost-benefits considerations.

Cosmopolitans respond that these considerations are part of the proportionality judgment and do not need to be duplicated in just cause. Since the former takes into account the magnitude of harm expected by war, it should serve to prohibit most wars over small rights violations without building this prohibition into just cause. However, in response, I would second May's (2008, 60) argument that "there is rudimentary proportionality consideration within the very idea of the principle of just cause." It does not make sense to treat a single murder or a group of non-basic rights violations as "cause for war" contingent on a subsequent proportionality judgment. They are not of a magnitude to provide just cause, such that intervention and its proportionality should not even be considered.

In my view, proportionality remains an independent criterion to test whether an individual war will be beneficial. Even a just cause of an in-principle sufficient magnitude could be disproportionate to go to war over due to circumstances that make likely costs outweigh benefits (for example, if it were expected to trigger a catastrophic nuclear war). However, it is a mistake to treat all rights violations as cause for war, subject to an additional judgment that fighting it would be proportionate. As discussed in Chapter One, proportionality is imprecise and easily misjudged, such that it provides little bar to war in itself. A low just cause threshold leaves the war decision to utilitarian calculation, which there is both utilitarian and principled reason to reject.

McMahan typifies the cosmopolitan view in his assertion that "just cause says nothing about considerations of scale or magnitude" (2005, 4). From his individualist perspective, war is viewed as a collection of forceful but proportionate responses to particular rights violations. Each action has its own just cause taking away the need for an overarching larger just cause. However, in war as we know it, it is not the case that once it starts, individuals are targeted narrowly because of their personal moral liability. War allows the killing of enemy combatants – typically members of the enemy armed forces – who have not necessarily committed any wrong and accept the killing of innocent bystanders through double effect more readily than we do outside of war because of the importance of ending the war and achieving the war cause. It also puts one's own combatants and non-combatants at risk to send them to war by inviting defensive responses by the enemy party. Because recourse to war uniquely sets in motion a series of resorts to force at a lower threshold than is normally permitted, including killing the innocent, the recourse to war requires a particularly serious cause.

Individualists and cosmopolitans sometimes admit this. Later on, in the article cited above, McMahan contradicts the magnitude-free view, acknowledging that since war "involves killing and maiming, typically on a large scale ... this gives considerations of scale a role in the concept of just cause. Only aims that are sufficiently serious and significant to justify killing can be just causes." In the next sentence, though, he adds, "Beyond this, however, considerations of scale are irrelevant to just cause" (2005, 11), attempting to reassert his magnitude-free cause thesis. However, having admitted that the cause must be proportionately large-scale, the qualification negates the thesis. That even such cosmopolitans recognize the need for large-scale injustices before one can go to war indicates the compelling reasonableness of the requirement.[17]

In terms of thresholds for intervention, above I mentioned Hudson's account wherein discrimination against women or ethnic and religious minorities can amount to an atrocity and, thus, a just cause for war. I think this strains the meaning of "atrocity" and justifies force too readily. Unequal treatment and violations of civil liberties are unjust and can rightly be criticized. However, it is generally not a cause for armed intervention. Hudson points out that discrimination can have material consequences. For example, in Afghanistan under the Taliban, the denial of civil liberties, including the right to education and employment, contributed to reduced life chances for women. Hudson is right to call this a problem of "human security." While I believe that these effects are convincing in showing that the policies in question are unjust and should be changed, I am skeptical about calling inequality combined with statistical differences in life chances an atrocity and cause for war. Although we can speak of these norms as committing a sort of violence, they do not justify the response of large-scale killing involved in war. In the United States, African Americans, Native Americans, and other racial minorities have a significantly lower life expectancy than white Americans, presumably due to the history of and continuing racial injustice. While this is indicative of a serious wrong, it is intuitively not a cause for war. In institutional injustices, the remedy is complex, responsibility is diffuse, many of the evils are unintended, and the victims may be partly complicit in continuing the oppressive system. Armed violence, particularly from external intervention, is not an appropriate response. Hudson's case for intervention in Afghanistan is strongest when she mentions the violence imposed by the Taliban in the form of rape, physical abuse, and brutal executions of women and men who were viewed to transgress its norms. If widespread, this is the sort of mass atrocity that justifies war. Without such physical assaults, I suggest that neither discriminatory laws nor divergent life expectancies, nor even their combination, justify armed intervention.

I have defended a threshold which I take to have intuitive plausibility, be justified by conceptual deontological arguments, parallel international law, and (as I shall argue in-depth in the next chapter) be desirable in terms of consequences. I turn to another influential argument for intervention.

4.7 Intervention versus revolution: the case for asymmetric standards

We saw that a central argument in favor of lowered bar of intervention is the analogy to the cause for revolution. Most would argue, following Locke, that the denial of rights, including basic liberties, social equality, and democratic political participation, is a just cause for revolution against one's government. Cosmopolitan hawks reason that if the denial of these rights can justify violence by the victims, it shows that the oppressive government is liable to attack and that an intervener can justifiably exercise this right on the victims' behalf. In his systematic inquiry, Dobos (2012) finds that the only valid asymmetry between intervention and revolution is that foreign interveners must meet hurdles of internal legitimacy (125–55); that is, they must justify the intervention to their home public. This, in turn, he argues, implies that humanitarian interveners may have to satisfy a standard of likely success that does not apply to those exercising self-defense (73–80).

I would defend the asymmetry between revolution and intervention, even beyond Dobos' internal legitimacy condition. First, there is intuitively an immediate right to self-defense within one's own homeland, which does not automatically extend to other agents. Pragmatic considerations reinforce this sense of a natural right. While the right could, in principle, be transferred to others by invitation by the victim, without such a request, other parties do not generally have a right to intervene paternalistically on their behalf. The victims are in a privileged epistemic position. They can know with relative certainty that their rights are being violated and judge whether a violent struggle to attempt to achieve these rights will be worthwhile. By contrast, potential foreign interveners have more difficulty knowing the seriousness of the oppression and the value of war to the victims. It is also difficult for interveners to know the potential for local forces to implement a new regime that would better protect human rights. Locals' direct knowledge of the relevant facts and values supports permitting their use of force at a lower threshold of injustice than that for external intervention.

Of course, rebel groups themselves could misrepresent the people on whose behalf they fight. They also might fight too readily based upon personal resentments or self-interested political aspirations. Most rebellions are unjust, violating just war criteria by responding to insufficiently large injustices, acting without right intention, disproportionately, and with little likelihood of success. Even the US Revolution has been argued to fall short on just cause ("taxation without representation") and last resort grounds (Roche 2019). Most contemplated revolutions should not be undertaken.

Nonetheless, it is not plausible to reject all rebellions for causes short of stopping atrocities. Local parties can sometimes reliably recognize and respond to systematic injustices at a lower threshold than mass atrocity. Moreover, it would be ineffective to ban rebellions – as insurgents, compared to states, are relatively unconstrained by international legal and moral norms. Typically, humanitarian intervention is only contemplated after an internal rebellion first signals large-scale resistance. This shows the primacy of rebellion and supports the relative ease of its justification.

Moreover, foreign intervention introduces systematic practical problems beyond those of national revolution. I address the problematic tendencies of intervention at length in Chapter Five. To briefly anticipate that discussion in the context of the comparison with revolution, external intervention tends to escalate conflict and be unstable in its gains. It is in inherent tension with local sovereignty. It also undermines a non-intervention norm that has served to limit armed conflict imperial domination around the globe. Internal insurrections do not undermine legal norms and cause destabilization in the same manner. Because of their immediate self-defense, epistemic privilege, greater harmonization with self-determination, and consistency with effective international norms, revolutions do not need to meet the same mass atrocity standard as humanitarian intervention.

A reasonable modified defense paradigm needs to offer more than Walzer and international law on the ethics of both civil wars and intervention in them. In terms of the justification of civil war, it seems clear that a party must have a serious cause to take up arms against its government, fighting and killing former compatriots over political control. At the same time, it seems clear that parties are sometimes justified in overthrowing local regimes. This may be because the government is colonial, as was the case in the Algerian and Latin American wars of liberation. Similarly, a war of secession to liberate a self-defined people might be justified, per Walzer (1977, 91–5). I would add the caveat that it must be that the people are treated unjustly within their current multi-national state and could effectively partition into a separate territory. Separate from secession, it seems that parties can justifiably overthrow an internally repressive government. From the French Revolution to the Cuban Revolution, and recently the Arab Spring revolutions in Tunisia, Egypt, and Yemen, rebellion in the face of injustice seems justifiable in principle.[18] As a general theory, following Locke and Orend, I would argue that systematic injustice in the form of any rights violations is necessary and sufficient for just cause for a revolution. The revolution must also satisfy last resort, proportionality, and legitimacy requirements.[19]

One may object that if revolution is justifiable resistance to injustice, then assistance to that revolutionary party would also be justified. However, blanket permission of assistance to rebels, even to rebels who have a just, rights-promoting cause, is dangerous. The lack of an international procedure for recognition of valid rebel causes leaves international assistance unilateral and disposed to arbitrariness and self-interested bias. International law has banned assistance to armed belligerents in a civil war. Gross (2015, 30–31) suggests that since the Additional Geneva Protocols recognize a right of insurgency against domination to further self-determination, this implies that external intervention on their part could be justified. However, a right revolution does not automatically entail a right to intervene on behalf of revolutions. If we do not beg the question in favor of intervention and revolution symmetry, pragmatism and legal precedent imply a higher bar for intervention.

The dubiousness of historic intervention supports retaining the non-intervention norm for civil wars. Even liberal states have bad track records of intervening for selfish political reasons rather than for human rights, with corresponding anti-humanitarian results. For example, US assistance to the Contras in their attempt to overthrow the Nicaraguan Sandinistas, like current Russian assistance to rebels in Ukraine, was undertaken in the name of human rights but actually undermined human security and democracy and was justly condemned. It is difficult to specify an effectively universalizable norm of intervention. This lawlessness is a reason to condemn interventions in civil conflicts without atrocity crimes.

Military assistance in the form of arms to justly fighting rebel groups or parties to a civil war seems less troubling than a foreign invasion, as there is less escalation and impingement on sovereignty, and the battle can only be won through participation by local forces. Nonetheless, the introduction of arms presents its own form of escalation, and armaments have a way of eventually being turned to use in conducting or supporting rights violations. Supporting rebels with arms shares many of the problems discussed above in sending armed forces, including mistakes and bias in selection and destabilization. There is reason that international law has condemned such support.

4.8 Defending the defense paradigm

In rejecting the HRP, I have implied that just war theory should retain or, if it has already abandoned it, return to the national defense paradigm, with exceptions in extreme cases, along the lines articulated by Walzer and the R2P. My central argument hinges on the problems with the logic and consequences of the human rights paradigm. However, insofar as I retain a modified legalist understanding of just cause, I need to answer the objections to this view laid out in section 4.2, which provide much of the motivation for the rise of the human rights paradigm. How can national defense be a just cause for war when protecting the individual rights that are national defense's *raison d'etre* are not? Shouldn't my mass atrocity threshold extend to national defense as well? And why aren't the ambiguities about what constitutes aggression and defense prohibitive?

My general line of argument is that retaining the defense paradigm better upholds human rights than a view that takes their violation to be a just cause for war. As I argued above, sovereignty generally protects human rights and well-being through the security and services it provides and through its furtherance of political self-determination. For the same reason that states which appear unjust but are not engaged in massive human rights violations should be considered legitimate, they also have the right of self-defense. Defense serves to protect the goods that are preserved by states, even flawed states.

I also argued that states which lack a legitimate government still have a right not to be interfered with from without. The illegitimate government, which lacks the right to rule over those people, nonetheless is justified in marshaling defense against an invasion. Recognizing a right of national defense has deterrent value since it means that would-be aggressors can expect resistance. To prevent unjust aggression, it may be necessary to permit questionable defense. At the same time, it would serve little good to prohibit unjust or illegitimate regimes from defending themselves since no regime is apt to recognize itself as such. It would be very difficult to get states to stop recognizing a right to defend themselves, whereas it follows naturally from mutual self-interest for states to reciprocally recognize a principle of non-interference.

As Walzer emphasizes, the defense paradigm is an intuitive framework for treating warfare, borrowing from our understanding of domestic conflicts between individual persons (1977, 58–66). Indeed, national defense and non-interference are the bedrock of international law, supported by both consensual agreement and the pragmatic value of effective norms. The development of a replacement paradigm would require new institutions governing the stable, effective, and just global enforcement of human rights. The world is far from such a state.

The theoretical possibility of bloodless invasion does not undermine self-defense as a just cause. Armed invasion almost inevitably leads to bloodshed. Moreover, even if the armed invader is interested in a bloodless takeover, it threatens the use of large-scale lethal force to coerce obedience. Such a threat itself is a form of serious violence that can justify defense. A party that issues such threats and coercion, violating recognized borders can be assumed willing to violate further rights, regardless of its promises. History gives reason to fear aggressors. This is the reason that domestic criminal law permits the use of force against home invaders who have not themselves used violence. The criminal trespasser has already violated rights and, by doing so, threatens more basic rights violations, which may become impossible to stop once they commence, thus *prima facie* triggering a right of defense (Lee 2012, 151–2).

Above, I noted that proponents of the human rights paradigm tend to hold that instead of talking about the rights of states as collectives, we should derive our ethics from the rights and responsibilities of individuals.

McMahan roots his theory of just cause in the same concept of individual liability to attack, which underpins his revisionist conception of *jus in bello* with its rejection of the moral equality of combatants:

> There is just cause for war when one group of people … is morally responsible for action that threatens to wrong or has already wronged other people in certain ways, and that makes the perpetrators liable to attack as a means of preventing the threatened wrongs. (2005, 11)

It is significant, however, that this refers to the responsibility and liability of a "group." Elsewhere I, among others, have questioned whether the use of military force between large groups, among whom aims and responsibility are varied, can be targeted according to individual liability.[20] Whatever one thinks of *in bello* actions, the initial *ad bellum* decision to fight in particular will necessarily be abstracted from individual liability. McMahan notes but sets aside the complicating factor that "there may be wrongs that are not entirely reducible to wrongs against individuals because they have a collective as their subject" (2005, 12). Not just the original crime but also the war to stop that group will be carried out by a collective. The scale of these endeavors means that we should not pretend that they are guided by individual liability to force. The *ad bellum* decision is about whether to enter a realm where people are targeted regardless of their particular liability, thus doing considerable but not precisely calculable harm to the non-liable. The imprecision of the moral calculations of war requires a general characterization of the circumstances in which the collective use of force is warranted.[21]

Perhaps the most consistent individualist account of military ethics is that of Rodin (2014), who rejects war upon groups in favor of policing actions responding to individual threats and crimes. Although I am largely sympathetic to this contingent pacifist view, I am inclined to think there are, and for the foreseeable future will be, rights violations of a magnitude that make it necessary to mobilize against states or other groups, i.e. to fight a war. Because of the moral messiness and complex risks of such mobilization, (1) it has to be determined by general types of situations which pose a large-scale threat to human rights, and (2) these circumstances should be exceptional to prevent the too ready recourse to the extreme means of war. I suggest, following Walzer, that national defense is one such circumstance and intervening to stop mass atrocities is another.

A norm of national defense does not have the negative consequences that a norm of intervention would. National defense generally fosters peace and stability. Defense suggests the possibility of returning to the status quo, living and letting live, while intervention implies a principle of ongoing struggle for control. Intervention is inherently destabilizing, leading to intractable problems without a clear end-point with likely success. In Chapter Five, I argue that humanitarian intervention as a principle is subject to misapplication and abuse. There is much less risk of this in national defense. This is in part because the violation of borders is relatively clear. States know whether or not they are under attack more readily than they know whether human rights violations in a foreign country are severe enough to justify intervention and whether this intervention is desired by the local population.[22] Admittedly, a norm of national defense can be abused or misapplied – as the US did in its war with Spain over Cuba after the misperception that the US Maine was sunk intentionally and with the Johnson administration's false claim that the North Vietnamese had attacked a US ship in the Gulf of Tonkin rationalizing war in Vietnam. However, the requirement of an actual attack, as opposed to unjust human rights violations, is precise enough to reasonably restrict the causes for war. Our ability to recognize such cases as clearly unjustified shows the meaningful content of the concept.

Lee's criticism of the difficulty of interpreting the defense paradigm is puzzling when used as an argument for preferring the human rights paradigm. What constitutes a sufficiently grave violation of rights is more ambiguous than what constitutes forced border crossing or interference with sovereignty. The national defense paradigm leads to seeming paradoxes because it makes specific recommendations for when war is and isn't justified. If the HRP seems to avoid this, it is because it makes no clear recommendations whatsoever. A case for war on human rights grounds might be asserted in most any conflict, sometimes by both sides, without being definitively refutable. The legalistic defense paradigm is relatively clear by comparison.

There will be borderline cases of defense, where arms or lesser force is used across borders, violating sovereignty, but without necessarily an intent to conquer or seize territory. To deal with these, a concept of national defense as just cause should recognize degrees of violation of sovereignty which trigger different degrees of response short of full-scale war, including calls for sanctions and reprisals in kind. Indeed, international law has developed reasonable precedents and principles for adjudicating these conflicts short of war. Embracing the defense paradigm does not require following Walzer in treating every armed border crossing or infringement of sovereignty as an act of full-scale aggression and just cause for war (1977, 52). Nor do we need to agree with *Just and Unjust Wars'* problematic assertions that in cases of disputed sovereignty, "the land follows the will of the people" (55) and that a national minority can secede at will (91–95). Stability around the globe requires a presumption in favor of the continuation of accepted boundaries when these are not the cause of large-scale human rights abuses. Additionally, a modified legalist paradigm need not uncritically accept all military responses to aggression as just. National defense, even when considered a just cause, must meet the other just war criteria, including last resort, proportionality, and discrimination if it is to be morally acceptable.

In regards to Lee's "time lag problem," it will not always be clear at what point a conqueror who could be justifiably ousted on the principle of defense becomes a new sovereign who is immune to attack on the same ground. However, these judgments can be reasonably made. Recent aggressors who are widely resisted and resented can still be overthrown. Former aggressors whose rule has become accepted acquire sovereignty rights. The state defense model has to make baseline assumptions about where state borders are and which governments are legitimate. The principle of *uti possidetis* ("as you possess"), stipulating the presumption of maintaining current borders, has long been recognized as the governing norm in international treaty and customary law and has served as an effective basis for making these decisions (Cassese 2005, 83-4). Law, utility, and common sense are mutually supporting here.

4.9 Conclusion

In conclusion, the objections to the defense paradigm are not only not fatal but also negated by a plurality of reasons that support its continued recognition as the standard of just cause. Since the national defense paradigm furthers human rights and well-being on balance, it is effective to think of defense of national borders as just causes and acts of cross-border aggression as lacking just cause. Injustices that do not cross borders are generally of a lower magnitude and do not justify war, while the effort to intervene across borders poses more threats to the international community and the rights of individuals. I have argued that sovereignty is not forfeited and intervention justified as readily as cosmopolitan interventionists assert. The presumptions in favor of self-determination and against violence, war's risk to lives of innocents without their consent, and the destabilizing effects of war all imply that intervention should be restricted to cases of mass atrocity and not lesser, much less any and all, human rights violations. In this chapter, I have emphasized conceptual and deontological reasons for rejecting the fashionable human rights paradigm in favor of a modified legalist paradigm. In the next, I discuss pragmatic considerations about the best way to protect human rights and welfare.[23]

Notes

1 I use this term from Steven Lee (2012, 121).
2 Theorists endorsing versions of the human rights paradigm include Lee (2012), Luban (1980a, b), Teson 1988, Frowe (2014), Buchanan (2010), Fabre (2012), Orend (2013), and Lango (2014), to name just some salient proponents.
3 "Individual rights (to life and liberty) underlie the most important judgments we make about war." (Walzer 1977, 54).
4 Rodin (2003) supports a variant of the human rights paradigm in which the use of force must be justified by actions to protect individual rights. However, his is

not a hawkish cosmopolitanism in that it tends to support contingent pacifism in which war is rejected in favor of individual acts of defense and policing.

5 See Holzgrefe (2003, 36–49) for a discussion of the legality of intervention under Charter Law. He notes that this prohibitionist interpretation was the standard view prior to 2000.

6 Bellamy and Luck (2018) refer to these as the four atrocity crimes. Others use the term "Mass atrocity crimes" to emphasize their unique nature as involving the large-scale killing of innocents.

7 Even before the R2P, some scholars argued that humanitarian catastrophes could be considered to fall under the Chapter VII threshold of "threats to international peace and security," insofar as human rights are understood as international principles. In the last fifty years, international law has come to recognize some principles as so basic to the concept of international order that they are per-emptory norms, or *jus cogens*, which are non-derogable through other treaties or state preference. The *jus cogens* norms have been understood to include the bans on genocide, slavery, and other crimes against humanity, as well as aggression. (Cassese 2005, 198–212).

8 Fabre (2012) makes a similar argument.

9 Fabre has a similarly strict cosmopolitan criterion for legitimacy, and skepti-cism about sovereignty rights against humanitarian intervention, but denies (reasonably, as I argue in Section 4) that loss of legitimacy alone justifies armed intervention (2012, 169–71).

10 Teson has an odd view of proportionality in which it is the actual consequences that determine the justification of war (Teson and Ven der Vossen 2017) . Thus, the Iraq War is now unjust but could become just if the consequences take a turn for the better. If a reckless intervention ends up by freak chance having beneficial consequences, it becomes justified. A more plausible view is that of Frowe (2014), in which it is the expected consequences that determine justifiability. Frowe concurs with Teson about the lack of significance of sovereignty and also jettisons the requirement of the consent of those to be rescued, making her view even more permissive.

11 See Chapter One.

12 I am influenced here by Richard Norman's (2006) argument that ambiguities about what constitutes human rights undermine the cosmopolitan justification of humanitarian intervention.

13 For a report of the worsening situation following the Arab Spring see Human Rights Watch (2019).

14 The imposition of risk on non-consenting parties is at the core of Bas Van Der Vossen's argument for a presumption against intervention (Teson and Van der Vossen 2017).

15 George Lucas (2014) makes this analogy to emphasize the moral complexity and difficulty of humanitarian intervention.

16 Even the earliest formulation of the R2P stated that, "Military intervention for human protection purposes must be regarded as an exceptional and extra-ordinary measure" (ICISS 32).

17 Lango's cosmopolitan conception of just cause contains a magnitude component in the form of proportionality between type of action contemplated and the magnitude of threat to human rights. With this conception he defends the pos-sibility of a low cost armed conflict to save the life of a single person (2014, 116–22). In addition to my arguments about the benefits of a magnitude threshold here, I take into account objections surrounding the possibility of low-cost interventions in Chapter Five.

18 I mentioned above that these revolutions mostly had bad outcomes. Because of their low likely success and disproportionately bad results they usually should not be undertaken.
19 The last, authority, requirement is the trickiest for revolution. I take this up in Chapter Nine.
20 For my critique of McMahan's revisionist *jus in bello* see Rocheleau (2010a and 2010b).
21 For further argument against individualist reductionism see Lazar (2014) and Emerton and Handfield (2014).
22 The case is somewhat more difficult for collective defense, i.e. going to the aid of another attacked state. The decision to do this may rely in part on trusting the word of the purported victim. However, I would argue that news and intelligence reports can more easily give a clearer picture of inter-state aggression than intra-state human rights abuses.
23 Despite his principled defense of the HRP, Lee (2012) anticipates my view when he suggests that we may nonetheless for practical purposes have "considerations of just cause be governed by a *prima facie* moral rule of thumb that takes as the fundamental value state sovereignty" (305).

References

Beitz, Charles. 1980. "Nonintervention and Communal Integrity." *Philosophy & Public Affairs* 9 (4): 385–91.
Bellamy, Alex, and Edward Luck. 2018. *The Responsibility to Protect: From Promise to Practice.* Cambridge: Polity Press.
Bohman, James. 2008. "War and Democracy." In *War: Essays in Political Philosophy*, edited by Larry May, 105–23. Cambridge: Cambridge University Press.
Buchanan, Allen. 2007. *Justice, Legitimacy and Self-Determination: Moral Foundations for International Law.* Oxford: Oxford University Press.
Buchanan, Allen. 2010. "Institutionalizing the Just War." In *Human Rights, Legitimacy, and the Use of Force*, edited by Allen Buchanan, 250–79. Oxford: Oxford University Press.
Cassese, Antonio. 2005. *International Law*, 2nd edition. Oxford: Oxford University Press.
Cohen, Jean. 2008. "Rethinking Human Rights, Democracy, and Sovereignty in the Age of Globalization." *Political Theory: An International Journal of Political Philosophy* 36 (4): 578–606.
Dobos, Ned. 2012. *Insurrection and Intervention.* Cambridge: Cambridge University Press.
Doppelt, Gerald. 1980. "Statism without Foundations." *Philosophy & Public Affairs* 9 (4): 398–403.
Emerton, Patrick, and Toby Handfield. 2014. "Understanding the Political Defensive Privilege." In *The Morality of Defensive War*, edited by Cecile Fabre and Seth Lazar, 40–65. Oxford: Oxford University Press.
Esfandiari, Haleh, and Kendra Heideman. 2015. "The Role and Status of Women after the Arab Spring Uprisings" IEMed Mediterranean Yearbook. Accessed November 5, 2020. https://www.iemed.org/observatori/arees-danalisi/arxius-adjunts/anuari/med.2015/IEMed%20Yearbook%202015_Panorama_WomenAfterArabUprisings_HalehEsfandiariKendraHeideman.pdf

Fabre, Cecile. 2012. *Cosmopolitan War*. Oxford: Oxford University Press.

Fassbender, Bardo. 2003. "Sovereignty and Constitutionalism in International Law." In *Sovereignty in Transition*, edited by Neil Walker, 115–43. New York: Hart Publishing.

Fotion, Nick. 2007. *War & Ethics: A New Just War Theory*. London: Continuum.

Frowe, Helen. 2014. "Judging Armed Humanitarian Intervention." In *The Ethics of Armed Humanitarian Intervention*, edited by Don Scheid, 95–112. Cambridge: Cambridge University Press.

Gross, Michael. 2015. *The Ethics of Insurgency: A Critical Guide to Just Guerrilla Warfare*. New York: Cambridge University Press.

Holzgrefe, J. L. 2003. "The Humanitarian Intervention Debate." In *Humanitarian Intervention: Ethical, Legal, and Political Dilemmas*, edited by J.L. Holzgrefe and Robert Keohane, 15–52. Cambridge: Cambridge University Press.

Hudson, Kimberly. 2009. *Justice, Intervention, and Force in International Relations: Reassessing Just War Theory in the 21st Century*. London and New York: Routledge.

Human Rights Watch. 2019. "World Report." Accessed November 5, 2020. https://www.hrw.org/world-report/2019/country-chapters/egypt.

International Commission on Intervention and State Sovereignty. 2001. *The Responsibility to Protect*. Ottawa: International Development Research Centre.

Lango, John. 2014. *The Ethics of Armed Conflict*. Edinburgh: Edinburgh University Press.

Lazar, Seth. 2014. "National Defence, Self-Defence and the Problem of Political Aggression." In *The Morality of Defensive War*, edited by Cecile Fabre and Seth Lazar, 11–39. Oxford: Oxford University Press.

Le Guin, Ursula K. 1971. *The Lathe of Heaven*. New York: Avon.

Lee, Steven. 2012. *Ethics and War*. Cambridge: Cambridge University Press.

Loughlin, Martin. 2016. "The Erosion of Sovereignty." *Netherlands Journal of Legal Philosophy*, 2016 (2): 57–81.

Luban, David. 1980a. "Just War and Human Rights." *Philosophy & Public Affairs* 9 (2): 160–81.

Luban, David. 1980b. "Romance of the Nation State." *Philosophy & Public Affairs* 9 (4): 392–97.

Lucas, George. 2014. "Revisiting Humanitarian Intervention: A 25 year Retrospective." In *The Ethics of Armed Humanitarian Intervention*, edited by Don Scheid, 26–45. Cambridge: Cambridge University Press.

May, Larry. 2008. "The Principle of Just Cause." In *War: Essays in Political Philosophy*, edited by Larry May, 49–66. Cambridge: Cambridge University Press.

McMahan, Jeff. 2005. "Just Cause for War." *Ethics & International Affairs* 19 (3): 1–21.

McMahan, Jeff. 2009. *Killing in War*. Oxford: Clarendon.

McMahan, Jeff. 2010. "Humanitarian Intervention, Consent, and Proportionality." In *Ethics and Humanity: Themes from the Philosophy of Jonathan Glover*, edited by N. Ann Davis, Richard Keshen, and Jeff McMahan, 44–73. Oxford: Oxford University Press.

Nathanson, Steven. 2010. *Terrorism and the Ethics of War*. New York: Cambridge University Press.

Naticchia, Chris. 2005. "Recognizing States and Governments." *Canadian Journal of Philosophy* 35 (1): 27–82.

Norman, Richard. 2006. "War, Humanitarian Intervention and Human Rights." In *The Ethics of War: Shared Problems in Different Traditions*, edited by Richard Sorabji and David Rodin, 191–207. Burlington, VT: Ashgate Publishing.

Orend, Brian. 2013. *The Morality of War*, 2nd edition. Buffalo, NY: Broadview Press.

Rawls, John. 1999. *The Law of Peoples*. Cambridge, MA: Harvard University Press.

Ricks, Thomas. 2007. *Fiasco: The American Military Adventure in Iraq*. New York: Penguin Books.

Roche, John D. 2019. "'Fear, Honor, and Interest': the Unjust Motivations and Outcomes of the Revolutionary War" In *America and the Just War Tradition*, edited by Mark. D. Hall and J. Daryl Charles, 50–73. Notre Dame, IN: University of Notre Dame Press.

Rocheleau, Jordy. 2010a. "License to Kill." *Review of Jeff McMahan's Killing in War. Radical Philosophy Review* 13 (2): 203–8.

Rocheleau, Jordy. 2010b. "Combatant Responsibility for Fighting in Unjust Wars: a Defense of a Limited Moral Equality of Soldiers." *Social Philosophy Today: The Public and The Private in the Twenty-First Century*, Volume 26, edited by John Rowan, 93–106. Charlottesville, VA: Philosophy Documentation Center.

Rodin, David. 2002. *War and Self-Defense*. Oxford: Oxford University Press.

Rodin, David. 2014. "Rethinking Responsibility to Protect." In *The Ethics of Armed Humanitarian Intervention*, edited by Don Scheid, 243–60. Cambridge: Cambridge University Press.

Runzo, Joseph, Nancy Martin, and Arvind Sharma, eds. 2003. *Human Rights and Responsibilities in the World Religions*. Oxford: Oneworld Press.

Shue, Henry. 2003. "Limiting Sovereignty." In *Humanitarian Intervention and International Relations*, edited by Jennifer Welsh, 11–28. Oxford: Oxford University Press.

Talisse, Robert. 2012. "Democratization and Just Cause." In *Morality, Jus Post Bellum, and International Law*, edited by Larry May and Andrew Forcehimes, 191–203. Cambridge: Cambridge University Press.

Teson, Fernando. 1988. *Humanitarian Intervention: An Inquiry into Law and Morality*. Dobbs Ferry, NY: Transnational Publishers.

Teson, Fernando. 2005. "Ending Tyranny in Iraq." *Ethics & International Affairs* 19 (2): 1–20.

Teson, Fernando, and Bas Van der Vossen. 2017. *Debating Humanitarian Intervention*. Oxford and New York: Oxford University Press.

United Nations. 1945. *Charter of the United Nations*. Accessed November 13, 2020. https://www.un.org/en/charter-united-nations/

United Nations General Assembly. 2005. Resolution 60/1. "2005 World Summit Outcome Document." Accessed October 13, 2020. https://www.un.org/en/development/desa/population/migration/generalassembly/docs/globalcompact/A_RES_60_1.pdf

Walzer, Michael. 1977. *Just and Unjust Wars*. New York, NY: Basic Books.

Walzer, Michael. 1980. "The Moral Standing of States: A Response to Four Critics." *Philosophy & Public Affairs* 9 (3): 209–29.

Walzer, Michael. 2000. "Preface to the 3rd Edition." In *Just and Unjust Wars*, 3rd edition, xi–xvi. New York: Basic Books.

Welsh, Jennifer. 2014. "Responsibility to Protect and the Language of Crimes." In *The Ethics of Armed Humanitarian Intervention*, edited by Don Scheid, 209–23. Cambridge: Cambridge University Press.

5 What is it good for? Consequences and the limits of war

5.1 Introduction

I have countered the cosmopolitan paradigm by arguing for a general presumption of sovereignty, even in cases where rights are being violated. Overriding self-determination and the norm of non-violence, including harm to the innocent, requires a large-scale, atrocity-level rights violation. For an intervention to be justifiable, it has to promote human rights to an extent that overrides its harms and rights infringements. On my pluralist view, if interventions at a lower threshold could be expected to have beneficial consequences, this would be good reason to accept them as a just cause. However, this chapter presents reasons to think that intervention typically undermines rather than advances human rights and well-being. One need not be a consequentialist to recognize that if intervention is likely to undermine the human rights values that are its cause, it should not be undertaken.[1] I will consider practical problems with armed intervention, including destabilization and escalation, error-prone judgment, para-doxical requirements for success, the erosion of international norms and cooperation, and imperialism. To complete the consequentialist argument, I respond to objections that intervention can be honed to be more effective, that generally negative consequences don't refute individual wars, that there is a continuum of small-scale wars with ordinary policing, and that non-interventionism in the face of human rights abuses is irresponsible.

5.2 Humanitarian intervention's unintended inhumane consequences

War inevitably causes death and destruction on a large scale. The attempt to overthrow, or otherwise forcibly intervene against, a rights-abusing regime frequently causes more violence – and thus loss of life and autonomy – than leaving the regime uninvaded. Although there may be civil violence even before intervention, outside force tends to escalate the conflict. Intervention brings a new party to the conflict with additional, typically heavier, weaponry, causing damage to infrastructure and collateral harm to non-combatants.

DOI: 10.4324/9781003105381-5

Armed intervention against a non-atrocity committing tyranny replaces repressive governance with mass mobilization, regular bombings, and the large-scale disruption of life. While Milosovic's Serbia, the Taliban's Afghanistan, Hussein's Iraq, Gaddafi's Libya were all abusive of human rights, loss of life was magnified by foreign intervention. It is doubtful that these interventions have saved as many lives as they have taken. In Afghanistan and Iraq, like Vietnam before, intervention fueled prolonged civil war.[2] In Somalia the US withdrew after not only suffering humiliating casualties but also having killed numerous civilians. In Kosovo, the NATO bombing campaign triggered the mass atrocity whose threat was its justification. Although not completely against intervention, Taylor Seybolt's (2007) study of humanitarian intervention in the 1990s cautions that proponents of intervention have "frequently underplayed the unintended consequences of intervention, for example in denying the role of bombing in triggering the massive expulsion of Kosovar Albanians" (94).[3]

The US and other liberal powers possess a combination precision-guided weaponry and codes of military conduct echoing the Geneva Conventions, which promise the ability fight a clean war, killing only enemy combatants liable to attack. The very concept of "humanitarian intervention" suggests that interveners will be benevolently focused on rescuing and protecting citizens. However, this is largely a myth (Fiala 2008, Chap. 8). In counter-insurgency operations, enemy combatants are dispersed through and difficult to separate from the population, leading to harm to non-combatants through accidental collateral damage and mistaken identification. Despite military emphasis on learning from the excesses of Vietnam, in Afghanistan and Iraq, torture, revenge killings, and other questionable killings, were common. Thomas Ricks describes how slow progress, the difficulty of gathering intelligence, and the US forces' lack of preparedness for the occupation and state-building mission contributed to a slide to abuse of the rights of prisoners and non-combatants in Iraq (2007, 214–97). Nor are US forces alone in being prone to violate rights in an intervention. UN peace-keepers have been found to have committed widespread sexual abuse in several interventions (United Nations 2020).

Interventions in the form of aerial bombing and support, like the campaigns against Serbia over Kosovo in 1999 and in Libya to help the rebels opposing Gaddafi in 2011, attempt to avoid the risks of on-ground counterinsurgency. However, such strategies come at the cost of relying on more destructive weaponry, and have more difficulty targeting sufficiently precisely to avoid collateral damage. Intervening forces have tended to put more emphasis on force protection than on protecting the victims of human rights violations whose rescue is their mission. This results both from the natural impulse of protecting one's comrades in arms and a strategic concern among intervening political leaders to avoid casualties that undermine home public support. In Kosovo, force protection concerns led to bombing from high altitudes to ensure zero US casualties while increasing the likelihood of civilian casualties on the ground. During the 2003 invasion of Iraq troop numbers were kept lower than security needs dictated (Ricks 2007, 41–3). In the initial occupation, the coalition authority and off-duty forces kept to the fortified green zone, removed from the Iraqi population (206–8). These decisions conserved American resources and lives in the short term, but are widely seen to have contributed to the ineffectiveness of the intervention (147–8). The presence of forces concerned to protect themselves from an insurgency endangers surrounding non-combatants who may be mistaken for rebels or killed unintentionally. The prioritization of force protection is not limited to US and NATO interventions; a report on UN peacekeeping missions from 2010–2013 found that the peacekeepers had suffered "no casualties as a result of placing themselves between attackers and their civilian victims" (Bellamy and Luck 2018, 153). Some ethicists have defended the prioritization of force protection over sparing civilians, reasoning that it is appropriate that a mission's beneficiaries bear its burdens (McMahan 2010; Øverland 2011) while others argue that combatants should take risks upon themselves to avoid civilian casualties (Walzer 1977, 154–9; Pattison 2014). In any case, foreseeable missions are likely to treat force protection as a greater priority than the protection of victims.

As introduced in Chapter Four, most casualties in modern interventions are non-combatants. Some are directly killed by interveners; others suffer and die as a result of the destruction of infrastructure, including utilities, transportation, and medical facilities; still others are killed by local parties exacting revenge for the intervention or seeking to gain power in the new vacuum. Although it is certainly true that other players share responsibility for these deaths, especially the killers in the last category, to the extent that these deaths would not happen without intervention and are foreseeable, the intervener bears at least partial responsibility. Even if killing innocents can be justified as unintended collateral harm through double effect, most agree that proportionality requires saving several lives to offset each death.[4]

In addition to direct destruction and killing, intervention causes desta-bilization. This occurs whether or not the intervention seeks regime change. If the perpetrators of rights violations are left in place there is likely to be no long-term improvement in human rights. For this reason, many interven-tionists recommend regime change as was done in Afghanistan, Iraq, and Libya. However, government overthrow creates a power vacuum that may be filled by parties as apt to abuse human rights as the former government. As I argued in Chapter Four, even rights-abusing governments frequently play a function in stabilizing society, preventing looting, ethnic conflict, and violence by disparate armed factions. The removal of Saddam Hussein in Iraq and Gaddafi in Libya caused all of these problems. The Kosovo in-tervention, while forcing out Serb forces involved in human rights viola-tions, left Kosovar forces to themselves engage in ethnic cleansing. Afghanistan and Libya have been embroiled in civil wars with human rights insecure years after their supposed liberation. Even armed interventions which do not seek regime change, by forcefully delivering aid and doing peacekeeping, create tension with local forces and may exacerbate conflict. For example, the intervention to protect the delivery of aid in Somalia resulted in conflict that was much more deadly for civilians than combatants (Seybolt 2007, 230–6). At the same time, at least one side's combatants if not both are presumably innocent. Even if we don't call their *in bello* killing murder, the foreseeable loss of their lives is an evil that counts against the decision to go to war.[5]

Even when armed intervention is followed up by occupation and state-building to preserve security and further reconstruction, as Orend (2013) recommends, it is difficult to achieve stability and human rights protection. Although authoritarian regimes in Iraq and Afghanistan were readily toppled, conflict, instability, and rights violations continue over 15 years later. While opponents of peace and human rights cannot achieve a decisive victory against the occupiers, they have been able to undermine the security necessary to make the interventions effective. Further exacerbating the problem, the military forces saddled with conducting interventions are not generally trained for policing and state-building operations. Commitment to ongoing state-building (as well as the legitimacy of continuing intervention) is apt to waiver, as interveners have political pressures and incentives to reduce forces and expenses. State-building will tend to be done "on the cheap," with a fraction of the recommended forces and economic support.[6] In order to supplement its armed forces in Iraq, the US outsourced significant security tasks to private firms. These groups, operating on a profit motive, have even less incentive and training in upholding human rights than intervening national forces.

Another reason that state-building is apt to fail is the lack of legitimacy of a government propped up by an intervener. Dependence on interveners may make the government appear (perhaps correctly) to be a puppet for foreign interests, lacking solidarity with the local people. A new government installed in power following repression may also exact revenge and continue sectarian conflict and rights violations. The people who have just emerged from repression and violent conflict may not have the will or political ability to resolve conflicts consistently with human rights. For example, the Iraqi government and army set up by the United States was more susceptible to Islamic State takeover of its territory then either the previous Hussein regime or the self-governed and un-reconstructed Kurdish region.

Experience and practical considerations lend further support to Walzer and Mill's argument that democracy cannot be instilled from without (Walzer 1977, 86–9). The attempt to forcibly establish democracy has been a "failure in every case after World War II except for small and very brief wars with Grenada and Panama," James Bohman (2008) finds.

> War is not an effective means to achieving democracy except in two specific cases: first when the war … reestablishes an already existing

democratic status quo; and, second, when the defeated non-democracy and not the democracy started the war in the first place (108).

It appears that stable, rights-respecting democracy requires virtues of citizenship and responsible government. If a democratic culture was lacking such that freedom fighters could not win on their own, it is unlikely to take hold under occupation. To the extent that the improvement relies on outside intervention – and the invasion would have been unnecessary if it didn't – it will tend to be unsustainable (Walzer 1977; Teson and Van der Vossen 2017). Moreover, reconstruction by a foreign power is apt to be poorly suited to the intervened-in state due to cultural difference and conflicting interests.

Interventions tend to face double binds. If they are small and short, eschewing regime change and state building, then they are apt to fail to establish sufficient stability, as seen in Somalia, Yugoslavia, and Libya. However if they are large, prolonged, and steering – as in Iraq and Afghanistan, they infringe heavily on self-determination, breed resistance, and create unsustainable costs. These paradoxes show that it is difficult for intervention to work.

Empirical studies confirm the difficulty of establishing basic security, much less a functioning democracy, after intervention. In his systematic study of dozens of foreign interventions in civil wars from 1900 to 2000, Patrick Regan (2000) finds that interventions rarely establish peace, only coming close to succeeding half the time when they take the side of the *government* in quelling opposition. Most recommended humanitarian interventions today would either side with insurgents or remain neutral, with little chance of decisively ending conflict. Regan adds that interventions have been most likely to succeed in stopping conflicts which had already become grave, with 10,000 or more casualties (Regan 2000). Seybolt's (2007) study of intervention in the 1990s according to a humanitarian metric reports mixed results, with operations saving lives in 9 of 17 cases. He concludes that while intervention has been effective in a variety of circumstances, using a range of strategies, "It is also clear that intervention has failed to save people on many occasions and involves a number of practical and ethical problems," and that "[h]umanitarian intervention does not offer a long-term solution to violent conflict" (280–1).

These results support a modified defense paradigm with exceptional intervention rather than a cosmopolitan one of regular intervention. This traditional view allows us to condemn the harmful interventions by the US in Vietnam, Somalia, Afghanistan and Iraq, and as well as those by the Russians in Afghanistan, Georgia, Chechnya, and Ukraine.[7]

5.3 Uncertainty and error-prone judgment in intervention

My argument against intervention has relied on its general tendencies rather than certainties. This means that there might be exceptions in which intervention to prevent human rights abuses short of atrocities might be worthwhile. However, interventions are characterized by uncertainty resulting from limited understanding and control of complex military and political situations. This "fog of war" undermines our ability to predict the humanitarian success of any contemplated intervention. Of course, some unforeseen consequences may be desirable. For example war industries may get an economic stimulus or medical advancements may be made in the course of treating new sorts of injuries. However, unintended consequences are likely to be negative. In their survey of the history of US wars, Kenneth Hagan and Ian Bickerton (2008) have documented that each has led to unforeseen negative consequences that tend to outweigh gains made.

When war planners are focused on their own cause, they tend to be overly optimistic about their likely success and benefits. Excessive military confidence is an instance of what psychologist Daniel Kahneman (2011, 250) calls the "planning fallacy," in which people tend to envision the best case scenario and end up overly optimistic in their prediction of outcomes. In practice, war generally does not go according to plans; as the strategist von Moltke said, "no battle plan survives contact with the enemy" (2012). Dominic Johnson (2004) theorizes that overconfidence may have been evolutionarily beneficial for earlier humans competing for resources with their unarmed fellows, but ends up harmful now that it is possessed by the leaders of heavily armed nations.

As Richard Werner argues, while self-deception is a risk in any decision, we are particularly prone to it in decisions to go to war. We construct narratives that confirm our initial orientation. Memory is selective, allowing a focus on facts which seem to confirm one's decision. All people, even experts, are poor predictors of the future and tend to imagine greater knowledge and control than they have. Leaders confirm their decisions by consulting like-minded advisors, lending seeming validation of judgments without critical examination (Werner 2013). As nations we easily come to demonize a scapegoat against whom war is being considered. Feeling fear and indignation, people are able to be talked into or talk themselves into a conviction that war is morally necessary and that their side, as the good side, will prevail. The psychological tendency to confirm biases strengthens when there is sense of a collective project which involves life and death struggle. Questioning one's own nation appears traitorous while believing in and supporting its cause provides a sense of shared purpose (Hedges 2002). Deception – such as the false reports of a North Vietnamese attack in the Gulf of Tonkin and of uranium, rocket tubes, and terrorist links in Iraq – can also be used to secure the support of a public willing to follow its leaders. The leaders themselves tend to engage in self-deception. Werner summarizes, "Cognitive biases and heuristics cause weak memories, poor predictions, deceptive over-confidence, and bias for our group and against others. At both the individual and the collective level they explain our self-deception when we go to war" (2013, 40).

Although ideally, responsible governments would engage in critical reflection before embarking on intervention, military decision-making tends to fall into what is known as "groupthink," where a group forms around a predetermined consensus and members are expected to fall line or be ostracized as disloyal (Recchia 2017, 60). The Washington Post recently exposed that despite almost 20 years of assertions by military and administrative officials that US state-building in Afghanistan is promising, it turns out that the military officials privately knew that the mission was not going well and believed the civil war with the Taliban to be unwinnable (Whitlock 2019). This gap between private reflections based on objective evidence and a rosy public discourse is an eerie echo of the revelation of the Vietnam Pentagon Papers (Ellsberg 2003). Myths about a clean war which spares the innocent further contributes to this self-deceptive rationalization of armed force (Fiala 2008).

Teson attempts to preempt the worry about mistaken interventions by arguing that "[i]f part of the definition of humanitarian intervention is that states do not abuse it, then it is difficult to resist the conclusion that the adoption of a rule following humanitarian intervention will have beneficial consequences" (1988, 104). The problem is that the decision whether to intervene must be made before the consequences are known. As Tony Coady retorts, "we need a vocabulary that doesn't assume success in advance" (2003, 276). Interventions which are expected to succeed are still apt not to do so.

Taken as a whole, these tendencies of mistaken *ad bellum* decision-making lend additional support to the case for a high threshold for intervention. Since the expected benefit of an intervention would have to be great to justify large-scale coercion and violence, including killing the innocent, most cases of human rights violations do not justify armed intervention. This supports a modified legalist paradigm.[8]

5.4 Less intervention or bigger, better intervention?

While I have argued that the poor results of recent interventions point toward a reemphasis on the non-intervention norm, cosmopolitans counter that we can learn from these problems to intervene more effectively. For example much of the work criticizing the Iraq invasion emphasizes the failure to plan for Iraq's reconstruction, with the implication that better planning could have made the invasion successful and valid (Ricks 2007; Pirnie and O'Connell 2008). Some just war theorists have begun to emphasize planning for effectiveness as a criterion for undertaking intervention.

Orend links his defense of humanitarian intervention to planning for successful state-building (2013, 215–46). He cites the examples of the post-World War II reconstruction of Germany and Japan to suggest the possibility of forcibly reforming authoritarian, rights-abusing governments into liberal democracies. While he admits that recent interventions have not gone well, he argues that they could if they followed best practices, a ten point "core blueprint for transforming defeated, rights-violating aggressor regimes into minimally just societies" (226). Among the guidelines are forgoing sanctions, purging much of the old regime, demilitarizing, revamping education, and revising the constitution to protect rights. These recommendations are relatively uncontroversial,[9] though in Iraq excessive demilitarization in the form of disbanding the army contributed to the lack of security, and the purging of Baath party affiliates gutted government of experienced administrators (Ricks 2007, 158–61). Orend's survey of state-building literature concludes that the two keys for successful state-building are "physical security and economic growth" (2013, 222). If security from violence and economic opportunity are provided, society will stabilize, support for institutions and peaceful solutions will grow, and a cycle of improving political and cultural bases for human rights protection can be established.

The recipe has salutary goals that are good starting points if one finds oneself engaged in statebuilding. However, the suggestion that this is a blueprint for successful intervention is implausible. Its insufficiency should be clear after fifteen years of attempting to implement it in Afghanistan and Iraq. The reconstruction pillars of security and economic development collapse upon analysis. The ubiquitous suggestion that the key to defeating an insurgency and ending abuses of human rights is to provide "security" is essentially circular. If one could consistently provide security, one could defeat an insurgency, but providing such security in turn requires stopping insurgent attacks. The recommendation is mostly an unhelpful tautology.

The suggestion that stability can be achieved through economic growth is similar. Economic growth requires a flourishing society, but a flourishing civil society requires economic growth. Some defenders of cosmopolitan interventionism suggest that if free market principles are just brought to poor countries, they will naturally raise themselves up. One of the most influential proponents of military nation building has been Thomas Barnett, whose talks based on his *Pentagon's New Map* (2004) and *Blueprint for Action* (2005) became essential briefings for those interested in retooling the military for nation-building missions. Barnett theorizes that the places which consistently violate human rights and threaten terrorism do so because they are disconnected from the world economy:

> Eradicating disconnectedness is the defining security task of our age, as well as a supreme moral cause in the cases of those who suffer it against their will. Just as important, however, by expanding the connectivity of globalization, we increase peace and prosperity planet-wide. (Barnett 2005, xvi)

He suggests that military force can restore connectivity to these states in "the Gap" (centered, he suggests, in the Middle East and Africa), the region non-integrated into pacifying globalization. Bringing "Gap" states into the free trade relations with the rest of the world will lead to prosperity and security. A similar argument that embracing free trade can lead to global development and peace was widely promoted by Thomas Friedman in the *Lexus and the Olive Tree* (1999) and *The World Is Flat* (2005). His New York Times opinion pieces combined the celebration of capitalist virtues with calls for the use of force overthrow Hussein's government in Iraq to democratize the Middle East (Friedman 2003).

However, it is not case that market liberalization provides a ready path to human rights. Free market principles have not been helpful in many developing economies. Economic liberalization has not been able to provide prosperity and stability in Afghanistan and Iraq. Encouragement of global connection, even with an influx of foreign aid, has failed to lead to great economic growth, particularly of a sort combined with broad access to employment that could give individuals commitment to the system and make them less susceptible to recruitment for continued warfare. Even in Ireland and India, which Friedman held up examples of the free market's promise to raise all boats, the benefits of laissez faire capitalism proved fleeting and Janus-faced. Their temporary economic booms proved to rest in large part on speculative investment. Ireland's economy crashed and India's is foundering on lack of infrastructure, massive inequality, and corruption. After economic liberalization, Eastern European states such as Russia, Hungary, Serbia, and Bulgaria devolved into systems of crony capitalism, where economic injustice mixes with a lack of democracy. The wealthy industrialist societies themselves face crises of worldwide recession, large-scale debt, growing inequality, and environmental degradation. Economic studies more thorough and less sanguine than those of Friedman have found that market liberalization has diminishing returns and leads to increasing inequality, which in turn undermines democracy (Piketty 2014; Deaton 2013; Krugman 2018) Other studies confirm the influence of complex historical factors in economic development (Diamond 1997; Banerjee and Duflo 2019). The idea that we have a model for economic growth that can be readily exported to failing and war-torn states lacks credibility.

What of the examples of the reconstruction of Germany and Japan? Why haven't the successes of the Marshall Plan been repeated in the 65 years since? Couldn't they be resurrected? In response, there are many ways in which these cases were unique and, in any case, vastly different from most of the societies in which there is a contemporary interest in state-building. A frequently noted point is that Germany and Japan were so utterly decimated after years of total war that there was no passion for resistance to the occupation. I take it that there no positive lesson to draw from this, as the repetition of such destruction would obviously be immoral as well as counter-productive according to most statebuilding theory. Aside from level of decimation, there are several other important disanologies. Japan and Germany were relatively culturally homogeneous, facilitating solidarity and relatively conflict-free reconstruction compared to the multi-national, religiously sectarian, or tribal situations found in the Middle Eastern and African countries under consideration for intervention. The Axis powers were also economically and educationally advanced, enabling their flourishing in the world economy. At the same time, the post-World War II period was one of great economic growth. Opportunities are less readily available in the competitive market of the Twenty-First Century. Moreover, the post-WWII Europe and US undertook their state-building with unprecedented dedication, borne of their experience of inadequate reconstruction after World War I leading to the disastrous rise of fascism and WWII. The Allies' motivation centered on self-defense more centrally than humanitarian beneficence. Finally, although there was considerable bitterness among Japanese and Germans for the brutality of the Allied War and the length of the occupation, this was tempered by significant admission of German and Japanese culpability for aggression and the lack of a history of colonial domination of those states and corresponding resentment. By contrast, in the global South, the site of current and likely future intervention, grievances over European and Christian colonialism against people of color, Arabs, and non-Christians, makes resistance to occupation inevitable. Orend's assertion that humanitarian rehabilitation "cannot be accused of creating a new generation of enemies, nor can it plausibly be seen as sowing the seeds of a second war," naively presupposes intervener purity and the ability of people in war-decimated states to appreciate the intentions of their first world occupiers.

Orend (2013, 185–246; 2014) contrasts his robust "rehabilitative" model for concluding wars with a "retributive" model that would give punitive sanctions while leaving a rights-abusing government in place (paradigmatically Germany after World War I and Iraq after the first Gulf war). However, the dichotomy between retribution from a distance and full-scale regime change with rehabilitative occupation is false. On one hand, as Walzer (1977, 121) notes, regime change and occupation are themselves understood as punitive. On the other hand, punitive sanctions are not the only alternative to complete overthrow and reconstruction. One can exercise containment and use pressure short of armed invasion and occupation to influence reform. For example South Africa and the formerly Communist Eastern Europe had successful revolutions following a degree of foreign pressure but no armed intervention. The idea that the world can only progress through armed intervention and occupation is overly restrictive in its vision, resulting in dangerously, unsustainably frequent war.

Orend's most specific recommendation for improving state-building is that interveners use greater numbers of peacekeeping troops: "The nation building research shows that you need about 20 soldiers per 1,000 residents to stabilize and secure post-war occupations" (2013, 227). At their high point during the surges in Iraq and Afghanistan, the US and its allies were far short of that number. To achieve Orend's recipe for success, the US and its allies would have to deploy at least *four times* as many troops as it did to

Iraq and Afghanistan! At the same time, Orend and other cosmopolitans recommend that intervention become more frequent. I take it that there is no chance of both of these occurring. Even with its too-small deployments, the ongoing wars have led to a steady stream of casualties and budget deficits in the US. Repeated tours of duty have taken a toll on communities, and the wars became increasingly politically unpopular at home. There is lack of willingness to continue expenditures even at the recent level of state-building, much less at the increased rates called for by cosmopolitan interventionism. Although increasing overall military spending, President Trump has pushed to reduce military intervention and foreign assistance. It is clear from the recent past and current trajectory that any interventions likely to occur will be done with fewer personnel and financial resources than experts recommend for likely success. Just war theory should make recommendations for the justice of actual interventions, not those presuming a world of infinite resources and perfect benevolence. Defenses of intervention by just war theorists will likely be taken by political actors and the public to apply to the real world and not be understood as merely hypothetical.

5.5 Selecting and fine-tuning for successful intervention

Contra Orend's call for larger-scale regime change operations, the most promising cases for intervention are smaller-scale operations. If one looks below the large-scale quagmires of Vietnam, Afghanistan, and Iraq, one can find cases of focused, less ambitious armed interventions, which have had success in ameliorating human rights without overwhelming negative side effects. These provide interventionists with a more plausible model for future humanitarian operations. Teson cites the relatively small-scale US interventions in Grenada in 1983 and Haiti in 1994 as successes. Alex Bellamy, one of the main philosophical proponents of the R2P, emphasizes the intervention by the Economic Community of West African States (ECOWAS) in Cote d'Ivoire (2010–2011) which forced the defeated incumbent president to step down, installed the democratically elected candidates and tamed a deadly civil conflict (Bellamy and Luck 2018, 175). A similar intervention has helped resolve conflict and restore democracy in the Gambia (Hartmann 2017). In Mali (2012–2015) the French intervened to help overturn the seizure of territory by atrocity-committing Islamist rebels. A recent return of US forces to Iraq and Syria helped defeat the Islamic State (or Isis), stopping the latter's run of egregious human rights abuses in its occupied territory, including attempted genocide against the Yazidi minority. In all of these cases outside armed force appears to have been reasonably successful at promoting human rights at acceptable costs (Bellamy and Luck 2018). The prospects for such successful interventions pose a challenge to my anti-interventionist view.

There are a few things to note in response. One point is that the situations are unusual in they involve relatively weak rights-abusers who had internal rivals, such that it was relatively easy for an intervention to tip the balance of power. The cases are also atypical in that they all involved intervening parties with connections to and personal stakes in the intervened-in state. The French intervention in Mali, though from a distance, was invited by the regional ECOWAS. Although the French and Americans may not have had particular cultural affiliations and legitimacy in Mali and Haiti, their respective colonial histories in these locations resulted in at least partial responsibility for the human rights abuses being perpetrated. For example in Haiti, General Cedras and other members of the coup that overthrew democratically elected Aristide were US-trained and had worked with the CIA to undermine the Aristide government. Classic successful interventions conducted by India in East Pakistan, Vietnam in Cambodia, and Tanzania in Uganda also had this feature of a neighbor with vested interests in stopping the genocide – indeed they were more defensive than humanitarian in motive. Because of the uniqueness of these situations, they should not be thought to express the general likely success and proportionality of foreign intervention to prevent any and all human rights abuses. At the same time, the locations cited here continued to experience conflict and instability after the intervention, so they do not contravene the thesis that is difficult to intervene decisively for the good. I would also note that the US assistance in subduing Isis in Iraq, protecting the Yazidis and others, was invited by the Iraqi government and hence was not an intervention of the sort of that is under dispute. Similarly, the Australian-led, UN-authorized mission which successfully stopped atrocity crimes in East Timor in 1999 was approved by the Indonesian government and thus not an intervention.

At the same time, I would note that in some of these cases the ongoing human rights abuses were probably sufficient to fall under the R2P standard of atrocity crimes, meeting my just cause threshold. Although several helped to restore democracy, it was the political violence surrounding the elections that justified armed intervention rather than the failure of democracy. Had there not been significant violence, intervention to promote democracy in the cases above would have not have been sufficiently important to justify armed intervention.

Bellamy and Luck's study of the implementation of the R2P finds that actions by various parties – including local political and military leaders, civil society forces, as well as humanitarian interveners – can make a difference. They conclude in favor of a "responsibility to try" to prevent atrocities (Bellamy and Luck 2018, 186–93). However, whether and what sort of intervention will work depends on features unique to each case and is unpredictable (174–77).[10] Notably, in the three successes they cite in addition to the Ivory Coast – Kenya (2007), Guinea (2009–2010), and Kyrgyzstan (2010) – atrocity prevention occurred without armed intervention, but rather through negotiations between local parties mediated by international agencies. While I have agreed that military intervention can be justified as a last resort in the case of atrocities, I would argue that the uncertainty of success in any intervention, combined with the presumption against using force and placing others at risk without their consent, suggests that intervention be prohibited in cases short of atrocity.[11]

Bellamy finds that atrocity crimes, once commenced, can typically only be stopped by regime change, whether by revolution or foreign intervention (2014, 168–9). It seems that once states have resorted to atrocities, they are desperate and beyond the reach of ordinary diplomacy. If this is true, it tends to lend additional support to viewing human rights abuses as just cause for war. However, I would point out that this thesis only holds for when the state has resorted to atrocities. When it is only guilty of lesser human rights abuses, such as political repression, diplomacy has a chance to work. Indeed Bellamy and Luck find that the most successful interventions all sought to prevent atrocities, while failed interventions lacked this aim (175). While cosmopolitan hawks would have us intervene "early" to prevent situations from becoming dire, this is a problematic inference. In the early days of repression before atrocities began or were imminent (whether in Syria, Rwanda, Kosovo, or Libya) it would have been premature to intervene.[12] More robust diplomacy may have worked while armed foreign intervention was likely to hasten the resort to atrocities. As Bellamy and Luck's own work on the R2P notes, most of the potential actions for more effective human rights protection involve non-military measures: "[i]n place after place, practice has shown the benefits of early engagement, before policy options are narrowed to a binary choice between coercive military intervention and looking the other way" (2018, 193). Also, three quarters of regime changes to end atrocities have been brought about by national civil wars or coups rather than foreign intervention (Bellamy 2014, 170). Combined with the burdens upon and reluctance of the international community to supply peace keeping forces, it makes sense to limit interventions to the most severe cases. Overall, I am in accord with Bellamy's conclusion that ethically "externally induced regime change will remain a rare and exceptional pathway to the protection of populations from genocide and mass atrocities" (2014, 186).

5.6 Targeted strikes, the sliding scale *jus ad bellum* alternative, and the war–policing boundary

The last section notes that beneficial interventions have tended to be smaller scale, eschewing regime change and state-building occupation. Some would argue that a range of armed interventions short of full-scale war are justifiable at a lower threshold of just cause. Such measures might include peacekeeping troops who seek to protect non-combatants – perhaps in a limited safe zone – without trying to defeat perpetrators. Some missions aim specifically to protect the delivery of humanitarian aid. There are also targeted strike missions that seek to capture or, more frequently, kill an individual or group which poses a threat. The raid that killed Osama bin

Laden is only the most famous. Since 2002, in addition to numerous Special Forces operations, the US has conducted hundreds, perhaps thousands, of targeted drone strikes (Duffy 2015, 412–25).[13] These have typically been against suspected terrorists and thus fall more under national defense than humanitarian intervention, but could be used to target local human rights violators as well. Rescue operations also frequently fall in the category of relatively minor interventions with a low cost-to-benefit ratio. The Israeli raid to free hostages held at the Entebbe airport in Uganda is probably the most famous. John Lango (2014, 121–2) cites as a hypothetically justifiable small targeted operation a raid to save the kidnapped journalist Daniel Pearl in 2002 before his captors beheaded him.

All of these missions have a restricted scope such that they are planned to have less harmful effects than a full-scale war in which there is struggle with an opposing collective in an effort to seize and control territory. Targeted operations can minimize risk to non-combatants and limit combatant casualties. They also tend to involve less political destabilization. It is plausible to think that such interventions could be justified at a lower threshold of injustice and threat to human rights than required for full-scale war. In this vein some cosmopolitan human rights theorists take full-scale war to require something like the mass atrocity standard, while holding out that smaller interventions can be undertaken at a proportionately lower threshold of rights violations. Such a just cause "sliding scale" has a common-sense rationality.

In defense of a high just cause threshold above, I argued that there is an element of proportionality in just cause. The proportionalization of just cause can be used to justify intervention on sliding scale grounds. For a utilitarian, whenever expected benefits outweigh expected costs, there is just cause (Sinnott-Armstrong 2007). For cosmopolitans like Teson, since any rights violation can provide the deontological basis for justifying force, the justification of intervention primarily rests upon proportionality. In a more nuanced deontological defense of the sliding scale, Lango (2014) argues that "There is a just cause when there is both a just goal and a justly correlative means. The just goal is preventing sufficiently grave violations of basic human rights, and the means of achieving it is a justly correlative planned course of military action" (124), where "just correlativity [is] between the scale of seriousness of the threatened harm and the amount of armed force used to stop it" (116). Proportionality of the armed means to the human rights protection end is thus determinative of just cause so that lower levels of force are more easily justified.

An additional cosmopolitan argument for a just cause sliding scale is based on the continuity of small-scale armed force with policing. Most of us who are skeptical about war accept the use of police force to interdict ordinary threats to human rights far short of mass atrocities. Since there would seem to be a continuous scale of armed force from ordinary policing to large-scale interstate war, it follows that some military measures in the low to middle range of force ought to be justifiable at a threshold below mass atrocity as well.

I reject a sliding scale of just cause and defend the relevance of the distinction between war and policing measures short of war. I will argue that the former requires the mass atrocity standard, and this should presumptively apply to borderline uses of force. In my view, measures short of war are justifiable at a lower threshold, including stopping less than atrocity-level human rights violations. I take as representative Orend's (2013, 3) definition of "war" as "widespread armed struggle between political communities over governance." This struggle over governance does not necessarily aim at regime change. Armed actions which significantly infringe a group's self-determination or physical rights constitute war. War involves a relatively large-scale use of force, both by definition and as a necessary means of resolving the conflict between groups involved. The law of armed conflict regulating war is particularly permissive regarding killing, as it allows the targeting of enemy combatants who do not pose an immediate threat and the collateral killing of non-combatants if proportionate in value to a military target (Luban 2008). These three qualities of war – struggle between groups, relatively large-scale force, and permissive norms of killing – all provide a marked contrast with policing, where norms are enforced and rights and welfare protected without a group struggle, without military-level arms use, and with lethal force only used as a last resort.[14] Because it involves killing on a large scale without typical legal human rights protections, and because it pits communities against one another, threatening the self-determination and security of each, there is a strong presumption against entering war. This is reflected in the prohibitive norms regarding *jus ad bellum*. It contrasts with the use of police force which, with its distinct mission, scale, and means, is justified much more readily and is thus accepted as a relatively ordinary feature of civil society.

While the paradigmatic difference between war and policing is clear enough, the question is about the nature of borderline cases. Some targeted or peacekeeping missions might engage in international violence with military force but do so with small enough force that it is similar in scale to police measures. Is it arbitrary to retain a strict high threshold for the former, or should a sliding scale continuum not be recognized?

In response, I argue that war actions, even when planned as small-scale, are problematic in ways that policing is not and that a fixed high threshold is required to prevent military action from becoming ordinary. First, acts of war are particularly dangerous and unpredictable insofar as they engage in a violent group struggle and threaten self-determination. Unwanted intervention is generally perceived as and responded to as aggression, regardless of its intent. Even targeted strikes and peacekeeping, if not consented to, risk escalation. The interveners may be pressed to defend themselves with increased firepower. Even if their initial plan was for mild force, they might be drawn into a larger fight with significant casualties to combatants and non-combatants. Although it narrowly avoided it, the bin Laden raid could have led to a conflict with the Pakistani security forces, which would have multiplied immediate casualties and caused a breakdown in American-Pakistani cooperation, and created international conflict. As happened in cases like Kosovo and Somalia, if a targeted use of force appears to be failing to achieve its goals, there is a tendency to adopt more widespread destructive means. Or, interveners may engage in mission creep, where having achieved some humanitarian goals, they adopt more ambitious goals up to and including destroying enemy forces and regime change.[15] This occurred in Libya, where the initial UNSC-authorized mission was to narrowly stop the massacre of civilians in Benghazi but was stretched by NATO forces to include bombing to ensure a rebel victory over Gaddafi's forces and the latter's removal from power. The actual war ended up being more problematic than the protective intervention, which was initially thought justified. As I argue in the next section, permissive intervention undermines international peace and stability. Policing within a legal framework lacks this destabilizing effect.

In practice, uses of force short of full-scale war have tended to exceed the bounds of police targeting and approach the relatively free fire of armed conflict. This has happened with targeted strikes; while secrecy and disputes cloud the evidence, they collectively have had thousands of casualties. They frequently eschew any attempt to apprehend and prosecute the alleged terrorist. In the killing of bin Laden, while the circumstances are disputed, some team members reported that the plan was to kill rather than capture and that bin Laden was shot while unarmed, without being called on to surrender. Targeted strikes are conducted without judicial review, much less a conviction in a court of law (Duffy 2015, 425–6). Thus, they amount to rights-violating extrajudicial executions or assassinations. Also, while there isn't much doubt about the guilt of bin Laden, other selected targets are more dubious. Targets have included individuals associated with terror groups but not clearly actively involved, including children. For example, after US citizen and radical cleric, Anwar al-Awlaki, was killed by drone, the US also assassinated his 16-year-old son (Duffy, 415). In some cases, the targets have been clearly mistaken, such as when the US acknowledged killing 24 Pakistani soldiers with one strike (415). Although narrowly targeted, small military operations tend to kill non-combatants along with the combatants or suspected criminals. Bomb attacks, including drone strikes, are especially apt to kill civilians. Even in the Special Operations bin Laden raid, five other people were killed in the attack. Small military actions have become a way for states, especially the US, to combine permissively high levels of military force with the permissively low just recourse threshold of policing. Such strikes are typically done secretly, without the formal declaration and public record that typically accrue to either war or policing.[16] This makes their ethical regulation difficult and makes their purported justification by a sliding scale of just cause especially dubious.

My proposal, then, is that all uses of force must either (a) meet the *ad bellum* criteria, including the atrocity standard of just cause as well as last resort, or (b) adhere to a policing ethic that upholds strict human rights standards, where nobody is targeted with force except as necessary to prevent killing, and borders are not crossed without consent. In cases where the rights of war do not clearly apply, the limits of police measures should be adopted. And, where human rights-respecting police restrictions are not going to be strictly used, the *ad bellum* threshold should be applied. This would prohibit the manipulation of the grey area between war policing to allow greater permissiveness and destructiveness. In addition to using means which preserve human rights, such missions should be legally vetted by the appropriate decision makers and reported for public accountability.

The deployment of peacekeepers against the will of the intervened-in state, like targeted strikes, is an act of war that requires the atrocity threshold of just cause. Traditionally, peacekeeping has been conceived as consented to by the warring parties (including non-state parties to civil wars), as well as committed to minimal force and impartiality or neutrality between belligerents (Bellamy and Williams 2010). So long as it is consented to and does not take the form of war-fighting and engagement in the struggle to defeat one of the parties, peacekeeping should not be understood as war and does not have to meet the atrocity and last resort thresholds. However, cosmopolitans would blur these boundaries and permit peacekeeping without consent or allow peacekeepers to take sides in the conflict.[17] This is intervention and constitutes war so that for deontological and consequentialist reasons described here, it should meet the *ad bellum* threshold. In cases of atrocity, peacekeeping intervention is justifiable. However, my discussion cautions against permitting intervention at a low threshold in the name of cosmopolitan peacekeeping.

David Rodin (2014) proposes a variant of the cosmopolitan justification of ready intervention by arguing that humanitarian intervention should not be conceived of as war but rather as protecting the rights of individuals. He would not have intervention try to defeat a group and determine control of the government. In this way, it is mostly like policing rather than war. However, by forcibly crossing borders, it will infringe the understood sovereignty of the local governing groups and perhaps the people in general. And, as it seeks to enforce individual rights, it is apt to become engaged in group struggles. Thus, I doubt that armed intervention can consistently be so un-warlike as to be readily justified.

Rescue missions are, in a sense, a combination of targeted strikes and human rights protection. Because they can be immediately life-saving while also avoiding the prolonged struggle with the opposing group, with all its incumbent harms, they promise to be relatively benign. Many of the rescue missions viewed as justified, like Israel's Entebbe action, could be considered acts of self-defense rather than intervention insofar as a state protects its own nationals. In Chapter Four, I explained why a norm of self-defense is more justifiable than one of intervention. One might think that quick rescue missions of foreign nationals could be readily justified without meeting the atrocity standard as well. For example, a NATO force might seek to rescue individuals to be executed for small crimes in repressive regimes. However, crossing borders military to interfere with a foreign power is problematic and apt to lead to an expanded conflict, killing more people than the life to be potentially saved and causing instability as an action and a general norm. For all these reasons, I would reject such military use as an ordinary cosmopolitan political instrument.

In some cases, a targeted strike may be justified as part of an overarching war against an enemy which has committed aggression or large-scale violations of human rights. In this vein, it could be argued that the US was at war with Al Qaeda, justifying the targeting of bin Laden as an enemy combatant.[18] I do not defend the strict legalist position that there cannot, by definition, be war with a non-state group. However, we should view skeptically the idea of a "war on terror" in which anyone labeled a terrorist can be targeted and killed. This lumps together many disparate actors who are not part of a single organization or engaged in a single overarching conflict. As other commentators have observed, a war on terror would be never-ending, boundless in time as well as space. Moreover, when borders are crossed to fight non-state actors, such armed intervention has to be justified to the state in whose territory it occurs. The latter must either give permission for the operation or be found to be culpably refusing to stop the atrocities or aggression rooted in its territory, justifying infringement of its sovereignty.[19]

On the condition that party A is at war with party B and individual X is an active combatant for party B and cannot be otherwise apprehended, and the territory Q in which X resides is consenting or complicit, then targeted killing of X by A is justifiable. However, if it is possible to respond to suspected terrorists by arrest and prosecution, these policing measures should be adopted. And, if individuals are surrounded or overwhelmed, as was the case with bin Laden, they should be given an opportunity to surrender, even if they were in principle combatants. This all follows from the presumption of employing a human rights-respecting policing means if possible.[20]

I have argued for the distinctive significance of 'war' as a type of conflict (a) requiring the *jus ad bellum* threshold to be satisfied before recourse and (b) allowing killing not permitted by human rights norms and standard policing. I must acknowledge that international law currently avoids the term 'war' in favor of 'armed conflict' (Cassese 2005, 399–434; Duffy 2015). Armed conflict, as defined there, does not have to be large-scale, especially if it is between states. Some armed conflict, then, is not war, even though all war is armed conflict. This is designed in part to extend restrictions and permissions to areas that fall short of war, where states are unwilling to employ police norms. A side effect, though, is to lend support to the idea that some armed conflict could be initiated without satisfying just cause and last resort. It is important to reassert the need for *ad bellum* criteria for all armed conflict as well as to recognize that cross-border uses of military force should adopt policing rules short of those typical armed conflicts when possible. In addition to the distinctiveness of war, my argument draws on the value of restrictive, relatively bright-line rules to avoid a dangerously manipulable gray zone between policing and war.

In sum, then, there are legitimate forms of force short of war, which do not need to meet the *ad bellum* threshold. However, these must be small and human rights-respecting, neither adopting warlike rules of engagement nor initiating military intervention or otherwise struggling over sovereignty, actions that require *ad bellum* justification. For all wars and measures that go beyond policing and have elements of war in their military means or sovereignty-infringing character, the *ad bellum* threshold, including the just cause threshold of national defense or intervention to stop atrocity apply.

5.7 The value of the non-intervention norm

To this point, I have argued against interventions to stop human rights abuses short of atrocity due to generally harmful consequences in combination with the presumptive norms of non-violence and self-determination. However, some counter that even if interventions are not generally wise, individual interventions to combat non-catastrophic human rights violations may nonetheless be desirable and potentially justifiable. As Allen Buchanan has argued, typical failures do not prove a type of action to be wrong in all cases. In his discussion of preventive war, he argues that the "bad practice objection" to this form of war because of its generally undesirable consequences "cannot show that any particular act of preventive war is wrong. The 'bad practice' argument is only plausible as an objection to the general acceptance of a principle allowing preventive war" (2007, 130). Buchanan subsequently extends this reasoning to defend "forcible democratization" in principle as well (2010, 264–72). We might anticipate that an intervention in a particular case short of atrocity would have a relatively good chance of succeeding and be welcomed by the vast majority of the local population, even if this isn't true of intervention in general. Why not, as Talisse does in his defense of democratization as just cause,[21] consider human rights violations a just cause for war that usually is not worth acting on but which can and should be acted upon in those unusual cases where it is thought beneficial? To complete the argument that a lower threshold of human rights violations does not amount to a just cause, it is necessary to explain why the non-intervention norm is valid even in such cases.

Another way of putting this objection is to argue that the concerns over war's consequences that I have developed in this chapter are taken into account under the other *jus ad bellum* criteria of proportionality and likely success. Thus, the argument goes, they do not need to be taken into account as part of the fundamental principle of just cause.

There are two general reasons why I adopt a rule utilitarian approach to just cause. First, for the reasons discussed in Section 5.3, states and other groups are likely to be mistaken in their judgments that an intervention is proportionate and likely to succeed. A principle of just cause must be able to guide conduct. This, in turn, implies that it should refer to identifiable classes of grave rights violation in which war is apt to be a justifiable response rather than stating that any rights violations are just causes, inviting agents to think of whether particular actions they are interested in undertaking can be justified. It is practical to rule out classes of dubious interventions as lacking a just cause rather than accepting them as *prima facie* having a just cause, contingent upon an assessment of proportionality, likely success, and last resort, as the cosmopolitan hawk suggests.

When impulses push toward excess, experience points toward general failure, and the consequences of false positives are disastrous, one should err on the side of non-intervention. A relatively clear and high bar is required to prevent its being dangerously misconstrued and misapplied. Short of strict pacifism, which is morally counterintuitive and difficult to demand, a norm permitting intervention in only the gravest cases is relatively clear and can effectively prevent the proliferation of mistaken interventions and international destabilization. If one errs in interpreting a requirement of massive human right abuse, the resulting intervention will at least respond to substantial rights abuse, whereas misapplying a requirement of "substantial rights abuse" would mean intervening in cases where rights abuse is relatively insubstantial and very likely to be outweighed by the harm of armed intervention.

Moreover, as in the justification of "paternalism" more generally, one must take into account that the coercive action employed for the other's welfare may fail to benefit him or her. When the action involved is one that frequently leads to negative consequences, this gives good reason for not undertaking this sort of action.

Although any norm might be misapplied and it would be an *ad hominem* to reject a norm because of its misapplication by one of its proponents, the extent to which proponents of the human rights paradigm have themselves misapplied their norm, such as by arguing for the dubious and disastrous intervention in Iraq, illustrates its susceptibility to misuse. In addition to his support of the invasion and occupation of Iraq, Teson defended the US support of the Nicaraguan Contras in their attempt to overthrow the Sandinista government against international criticism (although acknowledging problems with the Contras). Teson also favored the US intervention in Grenada that has subsequently been revealed to be completely without necessity, motivated by the desire to raise then-President Reagan's faltering popularity.[22] Luban supported the increased interventionism of the 1990s but then became a "chastened cosmopolitan" after seeing the problematic results of the Kosovo intervention (2002).[23] Cosmopolitan theorists were eager to support the intervention in Libya, which overstepped its authorization and has proven to have questionable long-term benefit. That its own proponents misapply the human rights paradigm demonstrates the danger of its acceptance as the justifying framework for the use of force.

To this point, I have emphasized the tendency to misapply an interventionist principle. Beyond this, there is a deeper principled and practical reason for rejecting such a norm. As many moral thinkers have argued, ethical conduct is determined by general principles rather than the consequences in individual cases. One acts as part of a community in which we are answerable to others, and our behavior can affect that of others. Following principles also shows a willingness to support global cooperation and respect the legal and moral norms of the international system. From state laws to norms against lying, cheating, and stealing, it is widely recognized that these should generally be observed even in cases in which we foresee marginal benefits in breaking them.

A norm of intervention to prevent moderate human rights violations permits others to follow the same norm. Other users may not apply it with good faith. Thus, the good faith violation can be an incitement to bad cases. For example, since the Iraq War, Russia has intervened in Crimea, and Iran in Iraq and Syria on the grounds of protecting human rights, pointing to the American invasion of Iraq as a precedent. This is particularly true in the international community where the number of actors is small; international conduct refers to precedents set by others and reciprocal principles of action in a way that is less common for individuals in society. Norms are revised according to the actions of other states. Indeed international law is formed in large part by custom.[24]

To adopt a norm of interventionism would involve either breaking or changing current international law. Legal change probably would have to be achieved through a period of illegal breach until customary norms are recognized to have shifted, which is what some cosmopolitan theorists recommend (Buchanan 2001; Hoag 2007). Yet, undertaking systematic violations of international law does considerable harm. Regular violations undermine the law's force, not just on the violated rule but regarding other laws as well. The value of the rule of law and international reciprocity are additional reasons to refrain from intervention even when one suspects its results would otherwise be salutary.

Of course, if Chapter VII of the UN Charter were rewritten or reinterpreted to specify a right of intervention into moderate cases of human rights abuse, then the objection of undermining the law would no longer apply. However, to explicitly permit humanitarian intervention would magnify the danger of abuse. To address this problem, Buchanan proposes an institutional framework to legitimize forcible democratization along with preventive wars, furthering cosmopolitan human rights (2010, 266–76). The new institution, along with more permissive norms than current international law, would establish an accountability regime for acts of preventive war and forcible democratization. The regime would review such interventions, "facilitate a critical, impartial evaluation of the evidence," and "attach a significant cost to a negative evaluation" (272). Making interveners accountable would dissuade them from intervening without good reason.

In reply, I would argue that this hypothetical condition in which interventions are rigorously vetted does little to justify intervention in the present. The prospect of getting the world powers to accept a revised law that would trigger automatic sanctions for their misuse of force is slim in the political world for the foreseeable future. Buchanan criticizes "contemporary just war theory [as] methodologically flawed because it is insufficiently empirical," and that thus "The integration of moral philosophy with institutional analysis is required" (2010, 252). Yet, Buchanan's own proposal is highly abstract and lacking in empirical basis; in particular, he does not describe the actual institutional structure he is proposing or how it could be achieved given current political forces. Nor does he discuss the activity and potential of actual institutions and norms, like the UNSC and the R2P, which seek to do some of what he proposes. Buchanan qualifies that he is not literally recommending new institutions or interventions, only arguing that hypothetically institutions and permissive norms should be adopted if feasible. Nonetheless, with his emphasis on criticizing restrictive just war theory and defending the in-principle justifiability of forcible democratization along with preventive war, he is generally understood as having provided cosmopolitan support for a lowered bar of intervention. In practice, politicians are apt to note that cosmopolitan theorists have defended intervention without noting the hypothetical conditions that render such proposals literally inapplicable to the actual world. To avoid ideological misapplication and to be relevant, just war theory needs to keep clear about what norms are valid in the actual world.

A new norm of intervention would be harmful in other ways besides encouraging mistakes and abuse and undermining law. A policy of intervention can worsen conflicts. By making states that view themselves as potential targets fearful, it encourages their militarization. This increased emphasis on security makes the state even less inclined to uphold human rights. For example, the interventions by the United States and other world powers in Iraq and Kosovo have had a motivating force in pushing Iran and North Korea to accelerate their nuclear programs, thinking that possession of a nuclear weapon would deter Western attack. The new cosmopolitan interventionism is undermining the security system established by the UN Charter and Nuclear Non-Proliferation Treaty, which has prevented world war and nuclear war since World War II (Burke 2005).

Antagonism toward states labeled as "rogues" undermines not only arms control but also other potential international cooperation on global issues. Reciprocity in formulating and observing international norms will be key to dealing systematically with other global challenges, including combatting climate change and other environmental problems, public health measures such as infectious disease response, the eradication of crushing poverty and debt and preventing international terrorism. A stated policy of armed intervention may thwart dialogue regarding diplomatic solutions to the human rights abuses that are the intervention's cause. That the aggressive attempt to promote humanitarian ideals can be self-defeating by undermining international security agreements has been a theme of the realist tradition in international relations. Thinkers including Hedley Bull (1977), Hans Morgenthau (1954), and Reinhold Niebuhr (2008), though concerned with justice and at odds with Hobbesian amoral nationalism, caution that liberal crusading is apt to backfire.

It is not only the potential target that is given perverse incentives by humanitarian intervention. The prospect of intervention incentivizes rebel groups to escalate conflicts and refuse to negotiate, inciting rights violations by the government to improve the likelihood of armed international assistance. Alan Kuperman (2008) has argued that humanitarian intervention creates a "moral hazard," akin to over-insurance, inviting dangerous revolutionary activity and discouraging the political compromise on which working solutions typically depend. Kuperman argues that the rebel groups in Bosnia and Herzegovina, Rwanda, Darfur, and Syria all adopted a strategy of inciting government brutality in order to draw foreign intervention, which was their best chance for victory against militarily superior government forces. The emerging norm of humanitarian intervention may be unintentionally motivating civil wars and rights violations of the sort that such intervention is aimed to stop.

Kuperman's moral hazard argument is not without critics. Bellamy and Williams (2012) reply that the timing of rebel actions suggests that they are responding to immediate government repression rather than trying to provoke massacres as an appeal to the international community. They also argue that since most of the movements that Kuperman mentions occurred prior to the international adoption of the R2P norm in 2001, they are not explained by the new norms of intervention. Kuperman might counter that the norm of humanitarian intervention had been developing in the 1990s before its codification in the R2P – so an incitement strategy was still potentially relevant. While I cannot adjudicate with precision between these interpretations, it seems likely that multiple factors influence a decision to take up arms, with hope for intervention combining with other contextual causes. For her part, Hudson (2009) accepts Kuperman's premise that current norms give an incentive to exacerbate conflict. However, she infers the opposite conclusion: it is because the bar for intervention is too high (75). If liberal states intervened earlier before mass atrocities, there would be no need to provoke massacres. However, I take it that intervening "early," before atrocities are committed or even imminent, calls for intervening too readily. It will mean exacerbating conflict which may not have worsened otherwise. It would also involve preventively attacking those who are not liable to attack, increasing instability, and simply not be feasible and sustainable for the international community. I would not, as Kuperman's argument suggests, prohibit all intervention to stop human rights abuses. Such an absolutist norm is implausible and overreacts to the possibility of bad incentives. A restrictive threshold can skeptically treat the requests of insurgents for assistance and has a bar high enough that most parties would not set out to provoke it.

A norm of ready intervention is also likely to overstretch the resources of the international community. Most military interventions have depended on the vast military resources of the United States, which outspends the next ten military powers combined. There is economic and political pressure to cut US military expenses. The maintenance of technological superiority, continual overseas deployments, and the ability to fight more than one war simultaneously is costly. The Iraq War alone has cost trillions of dollars (Stiglitz and Bilmes 2008). Above I criticized Orend's suggestion that human rights could be secured by significant expansion of resources dedicated to intervention. Here I emphasize that a policy of readily undertaking intervention would further exacerbate this over-extension and prevent the world community from being able to act in the case of actual humanitarian disasters. For example, the ill-fated mission in Somalia contributed to the US failure to act in Rwanda, and recent commitments in Iraq and Afghanistan probably influenced refusal to intervene in Darfur. That the intervention in Libya exceeded its mandate and has had dubious consequences may make it hard to intervene to stop future atrocities, including that in Syria. Maintaining a limited policy of intervention would assist in ensuring that forces are available to stop the worst cases. There are, of course, calls for democracies other than the US, especially in Europe and Asia, to invest more in their military forces to expand the global capacity for intervention. However, liberal states such as Germany, England, France, Japan, Australia, and Brazil also face economic challenges and have difficulty sustaining political support for their status quo military commitments, much less increasing them.

In practice, a policy of intervention to stop any and all human rights violations would lead to even greater inconsistency in intervention than we see in the present, permitting intervention in less dire cases and taking away resources for intervening in the direst cases. Although inconsistency does not refute intervention, either in general or in particular cases, it does entail a less than optimal, if not unjustifiable, use of resources and, hence, a moral failure. These inconsistencies reveal non-moral motives for the use of military power under the rubric of humanitarianism, and this, in turn, undermines their legitimacy.

5.8 Imperialism and cosmopolitan interventionism's legitimation problem

The biased motives and inconsistent justice in selective intervention relate to another general problem for intervention's legitimacy and effectiveness.

In practice, cosmopolitan intervention will be undertaken by the relatively wealthy, white, predominantly Christian Western and Northern states, while the interventions will occur against the relatively poor, non-white, and non-Christian states in the global South. Insofar as it exemplifies and furthers the power of the North over the South, cosmopolitan interventionism can be considered imperialist.

This exercise of power is both a result and continuation of the domination of the third world by the first world, beginning with the naked conquest and enslavement of Africans, Asians, and Native Americans. Later 19th and 20th Century imperialism was justified as a "white man's burden," in Kipling's famous terms, where exploitation was combined with paternalistic humanitarianism. European colonialism in Africa and US colonialism in Latin America and Asia has featured a combination of efforts to reform inhumane local practices at odds with Western norms while also implementing exploitative economic regimes, which are themselves brutal. Colonial domination results in a loss of self-determination and lingering resentment both towards the colonizers and any who received favorable treatment under them. Notoriously, the Rwandan genocide had roots in the ethnic distinctions put to discriminatory use by the Belgians and the subsequent colonial interests of the French. The US, for its part, had supported (and continues to support) authoritarian regimes throughout Asia, Africa, the Middle East, and Latin America, including both Saddam Hussein and the Taliban, before they became enemies (Johnson 2004). Conflicts over diamonds, oil, gas, and other resources surround many of the places where human rights are abused. International parties have economic stakes in the access to these resources and the flow of goods and are unlikely to be purely humanitarian in their intervention decisions. The flawed humanitarianism of a globalization strategy built around market liberalization, discussed in Section 5.4 above, also exacerbates existing inequalities. As Richard Peterson argues, "In general, occupation inevitably brings the occupied society more tightly into the web of relations and processes that characterize the contemporary dynamics of the market, communications, and military power. As a result, the occupied are subject to a particular pattern of development" (2006, 217). Formal democratic rights of elections and rule by representatives all operate within the constraints of these institutions and the pressures of the occupying powers.

Cosmopolitans may respond that interventions guided by human rights promotion either are not imperialist or are a justified form of imperialism. In his reply to a charge of imperialism from Terry Nardin (2005), Teson suggests both of these responses, ultimately embracing the title of "humanitarian imperialist" as a compliment. However, it would be arrogant or naïve to think that we are now ready for benevolent imperialism. I have already discussed the potential for a norm of intervention to be abused. Interventionism effectively gives Western powers authority through the world that can be expected to be harnessed for self-serving as well as humanitarian reasons. Regardless of its initial intent, intervention is apt to aggravate global inequalities even as it provides a legitimizing excuse.

The concern to project power and exert global influence can be understood as a central theme of the United States' "war on terrorism" and occupations of Afghanistan and Iraq. Teson's defense of the Iraq War as a humanitarian intervention embraces the Bush Administration's "grand strategy" of achieving US security by "promoting liberal reforms in the Middle East and, indeed, the entire world" (2005, 11).[25] The blending of permissive proposals for intervention with the prevention of terrorism and protection of US economic and security interests is not unique to Teson. Burke (2005) points out that this conflation is a common theme of new internationalists, including Michael Ignatieff, Anne Marie Slaughter, and Jean Bethke Elshtain. US interests, rather than pure humanitarianism, can be predicted to guide its interventions and nation-building. Indeed this is exactly what has happened in Iraq, with the safety and security of the population taking a back seat.

Even if, counterfactually, interventions in the name of human rights were entirely benevolent in intent, one cannot expect third-world states to accept these uses of power as legitimate. Given the history of exploitation and human rights abuses committed by the wealthy North against the South, to proclaim a right to govern on paternalistic grounds is highly dubious. Members of the global South are understandably skeptical about interventionism. Mohammed Ayoob argues that even more dangerous than directly negative consequences of humanitarian interventions is that it threatens to "erode the legitimacy of an international society that for the first time has become truly global in character." (2002, 81–2, 63–6). The UN Charter's emphasis on self-determination and sovereign equality had advanced equal respect for peoples of the global South. While self-determination sometimes shields internal rights violations at odds with respect for human dignity cosmopolitans' claimed right and responsibility of intervention to enforce human rights norms can itself signal disrespect. The legitimacy deficit of humanitarian interveners can be seen in the trajectory of the International Criminal Court. The ICC initially had widespread support in Africa, but following its overwhelming use to prosecute African leaders for crimes, African states and their citizens have become skeptical of the court, and several African countries have withdrawn from the institution (Between Africa and ICC [editorial] 2016).[26] Since intervention, especially unilateral but even with UNSC authorization, lacks the juridical procedures of a court and involves a deeper threat to self-determination than the punishment of individual criminals, it is especially likely to be seen as biased. The R2P was approved in 2001 with global support, including from the South, but interventions in less than atrocity level cases, on cosmopolitan grounds, without regional validity, can be expected to create widespread opposition and diminished legitimacy.

Power must be justified to those over whom it is wielded. The Western powers who would be called on to implement cosmopolitan human rights norms are lacking in moral credibility. If there were evidence that the Western states had the will and ability to police the world in a human rights-preserving way, and if this were seen to garner wide support amongst people worldwide, then there would be good reason to accept cosmopolitan inter-ventionist view of just cause. As things stand for the foreseeable future, an international human rights enforcement regime established by a group lacking credibility and potentially exacerbating their relative advantages through its implementation is lacking in legitimacy.[27]

These foreseeable deleterious effects of a norm permitting ready inter-vention combine with the likely harm of any particular intervention and the principle of self-determination to make an overwhelming case for the rejection of the human rights paradigm of just cause.

5.9 The objection of inaction in the face of humanitarian abuses

Much of the skepticism about a norm of non-intervention centers on the prospect of genocides like Rwanda and Darfur. In such cases, it is the failure to intervene, rather than the act of waging war, that is unjustifiable, as reflected in the too-often repeated moral cry, "Never again!" My rejection of human rights enforcement as a just cause faces the objection that it will allow human rights abuses to remain unchecked and that inaction in the face of abuses is immoral.

In response, first, rejection of cosmopolitanism interventionism to stop any and all rights violations is compatible with exceptional intervention to stop atrocities. Unlike strict legalism, my modified variant recognizes the genocides in Rwanda and Darfur as providing clear just cause for inter-vention.[28] To provide a counterexample to my moderate anti-interven-tionism, one would have to cite cases of less than catastrophic rights abuses in which intervention was nonetheless an obvious good and argue that there is defensible, plausible norm which would permit intervention in such cases without also leading to excessive intervention.

Second, for reasons I explored earlier in the chapter, I am skeptical about the existence of such examples and even more skeptical about the adoption of a norm of intervention in such cases. By analyzing the effects of inter-vention and its inherent dilemmas, I sought to show that interventions typically cause greater human rights problems than they avert, while the justification of war would require an expectation of significant human rights improvement. Thus, my second response to the objection that anti-interventionism permits human rights abuse is that intervention in non-catastrophic cases will tend to violate more rights than it protects. Third, the rejection of war to end human rights abuses short of atrocity does not re-quire inaction. After exploring these second and third points in more detail, I end by explaining why these limits of intervention do not contradict my permission of intervention under the first.

The restriction of intervention to atrocity-level cases does not imply inaction in the face of serious, but sub-atrocity, rights violations. On the contrary, it is important to remember that human rights violations can be criticized and resisted with methods besides armed force. Humanitarian assistance may be offered to improve the conditions of refugees and endangered civilians. Without armed intervention and forced regime change, safe havens across borders can be established and defended without the magnitude of violence and complexity of armed intervention and war. The R2P calls for preventive diplomacy before armed intervention is considered. Bellamy and Luck's (2018, 175) study of its effective implementation finds that much atrocity prevention, including three out of four of their paradigmatically successful cases, involved diplomatic but not armed intervention. International norms can be backed up with pressure to comply short of armed enforcement (140–63). International criticism, sometimes called "jawboning," and exclusion of parties from the benefits of cooperation, or "outcasting," can pressure states to change their human rights policies (Hathaway and Shapiro 2014). Diplomacy can be backed by economic carrots and sticks – such as awarding or withholding assistance, trade fellowships, and security partnerships. Studies have found that non-military assistance – e.g. disaster relief, medical aid, and infrastructure projects – can save lives much more easily and less expensively than military intervention (Valentino 2011).

My support of interventions to prevent genocides while rejecting most humanitarian interventions raises questions of consistency. If interventions tend to fail, exacerbating human rights violations, and if there are alternatives other than military intervention that work better, why should military force not be eschewed in cases of atrocity crimes as well? The reason is that the human rights violations in these cases are of such a magnitude that armed intervention is unlikely to make it worse and is very likely to beneficially stop or curb an atrocity. The costs of continuing measures short of armed force may be too great or their likely effects too slow, such that armed intervention to stop atrocities is necessary and, thus, justifiable. At the same time, the norm of intervention to prevent mass atrocities avoids the harms and destabilizing effects of a norm of intervention to promote democracy and stop less egregious human rights violations. The norm of atrocity prevention is relatively clear and does not lend itself to rationalizing any intervention that a party is inclined to take out of self-interest or irrational suspicion. As I argued earlier, if misapplied, atrocity prevention might be stretched to justify missions to prevent serious human rights violations short of atrocity. By contrast, a norm of intervention to stop any serious rights violations could easily be misapplied to justify attacks in non-dire situations. The latter, such as the invasion of Iraq, are apt to do more harm than good. These considerations suggest that utility is maximized by retaining a modified legalist paradigm with intervention only to stop atrocities as opposed to alternative norms.

Additionally, with the R2P, the Convention on Genocide, and Chapter VII stipulations regarding international intervention, a right to intervene to stop atrocities is legal, while interventions for other cosmopolitan causes are not. Thus, limited intervention upholds international law and global reciprocity. By contrast, an effort to declare a general just cause of intervention to promote human rights would be destabilizing by either breaking current law or discarding the norm of non-intervention which has effectively limited international warfare.[29]

5.10 Conclusion

I have argued that intervention tends to have negative consequences. Individual interventions have worse than expected results due to over-confidence in the fog of war, as well as tendencies toward escalation and long-term destabilization. The normalization of intervention has even more untoward effects as it invites abuse and miscalculation, contributes to a spiraling incentive toward militarization, weakens norms against the use of force, and involves delegitimizing imperialism. Efforts to improve intervention by a theory of thoroughgoing state-building or targeted minor force tailored to circumstances both are apt to do more harm than good despite principled intentions.

These considerations combine with the conceptual and deontological arguments from Chapter Four to support the non-intervention norm, with exceptions for mass atrocities on the one hand and limited police operations short of war on the other. The promotion of on-balance-utility is not sufficient to justify the violent infringement of sovereignty. At the same time, in practice, humanitarian intervention tends to undermine the very goods of human rights and well-being that are supposed to be its purpose.

Notes

1 For example, Teson agrees that "war is justified only if the damage it causes is not excessive," such that "proportionality is a necessary condition for … intervention" (Teson and Van der Vossen 2017, 98). However, he denies that intervention generally has negative consequences that make it presumptively wrong (98–128).
2 Admittedly, the Iraq and Afghanistan interventions, like that in Vietnam, were primarily intended as US defense rather than as humanitarian interventions. However, some commentators supported each as furthering human rights, and the administration listed humanitarian grievances as a war rationale, among others. In both Afghanistan and Iraq, human rights promotion became the main rationale for continuing the interventions to reconstruct the states. Since state-building and human rights improvement were extensively but unsuccessfully attempted in each case, they illustrate the limits of promoting human rights through war. I would also note that the prospect of an intervener's national security interests mingling with humanitarian goals is likely to be relevant in future cases of humanitarian intervention as well.

3 As Seybolt notes, NATO claims that the Serbs were planning their campaign of expulsion in any event.

4 For example, Van Der Vossen suggests that proportionality might be saving three to five lives per person killed (Teson and Van der Vossen 2017, 158).

5 See the discussion in Chapter One on the relevance of combatant casualties to *ad bellum* proportionality.

6 In Section 5.4 below I discuss further the contrast between the recommendations of Orend and state-building literature and the sorts of interventions that are likely to occur.

7 Like the Iraq, Afghanistan, and Vietnam cases, these Russian interventions were not primarily humanitarian. However, they were proposed as such by the interveners, and cosmopolitan interventionists tend to accept that interventions can have mixed motives. They might not welcome the Russian interventions but would have trouble decisively condemning them insofar as they sought to stop some rights violations.

8 Below I take up the objection that the consequentialist concerns I am listing are considerations of proportionality but not matters affecting when humanitarian intervention is a just cause.

9 For my state-building recommendations, see Rocheleau (2008), which draws on Orend's (2002) earlier work on the subject.

10 For example, although one might think that early warning of an impending crisis or UNSC authorization would make an intervention more likely to succeed, they find a negative correlation between these elements and success. This may be because situations that evoke an early warning and Security Council resolution do so because of entrenched difficulty (instead of, implausibly, becoming difficult because of the warning and authorization). Seybolt (2007, 267–81) reaches a similar conclusion about the unpredictability of success.

11 In inferring cautiousness from uncertainty, my view accords with that of Van Der Vossen (Teson and Van der Vossen 2017).

12 Regan's (2000) finding that interventions are most likely to succeed when the conflict is severe is relevant here.

13 Because of secrecy surrounding the program, the precise numbers are not known.

14 This is not to say that police force is unproblematic and cannot be excessive. As I write this, there are protests calling for both greater restrictions in the use of police force, correction for racial bias, and the replacement of armed police with less violence-prone sorts of community relations and interventions. Among the criticisms of the police is their tendency to mimic armed forces. My preference for police force over war is compatible with also recommending measures short of armed policing when possible. A thoroughgoing ethic of policing is a pressing issue which is beyond the scope of the present work.

15 In Chapter Eight I argue that new war missions, as well as changed circumstances, require new testing according to *ad bellum* principles, including just cause. If this were done consistently and appropriately, it would avoid the escalation problem. However, since war has the tendency toward escalation in defiance of rational planning, this is reason to avoid embarking on war in less than grave circumstances.

16 The Trump Administration rescinded an Obama policy of publicizing civilian deaths in drone attacks. It also rescinded Obama's stated restriction of the method to "high value" targets. What was already a permissive killing regime has become further unbound and escalated (New York Times editorial 2019).

17 Bellamy and Williams (2010) do not explicitly argue against the consent requirement but leave consent out of their definition of peacekeeping (18) and list the insistence on consent by the parties as a restrictive factor in traditional

peacekeeping (191). Their replacement of the traditional concept of neutrality with one of impartiality is potentially positive in that it implies that peacekeepers can use force against one of the sides that is engaged in wrongdoing, for example, stopping rights violations or seizing weapons. However, if the side–taking becomes a sustained effort to defeat a force rather than prevent immediate harm, then, according to my argument, it would become war and require a higher threshold (and a legal mandate) for its prosecution.

18 One might question whether bin Laden was still sufficiently active to be a combatant. The US asserted that he was still involved in planning Al Qaeda attacks.

19 Some have theorized that Pakistan may have privately consented to the bin Laden operation, although it publicly objected.

20 Shannon Brandt Ford (2013) defends a middle level of armed force between war and policing against the criticism of Tony Coady. Ford argues that this new area will help moderate some actions short of war. As is clear, my view accords more with the views of Coady and others who would apply the law enforcement model insofar as possible and keep acts of war minimized. For an exploration of targeted killing, with different views about the relevance of war as opposed to law enforcement models, see the contributions to Finkelstein, Ohlin, and Altman (2012).

21 See Chapter Four.

22 Robert Gallucci, former Assistant Secretary of State for Political-Military Affairs and member of Reagan's national security team, reported that the US administration refused an offer to negotiate the transfer of the medical students whose safety was supposedly the primary reason for US intervention. This suggests a clear failure of last resort and probably of just cause (Gallucci 2003).

23 Interestingly, Luban supported intervention on the side of the Sandinistas, the opposite of the pro-Contra action recommended by Teson and taken by the US.

24 On the legitimacy of international law, see Franck (1988). On issues regarding the adjudication between the norm of self-determination and the legality and morality of intervention, see Holzgrefe and Keohane (2003).

25 Teson goes on to deny that the grand strategy behind the US interventions should be seen as intended, saying instead it functions as a mere motive. I criticized Teson's use of the distinction between motive and intent in Chapter One. In the Iraq War, self-interested motives were mirrored by a self-interested plan for intervention. It would be more accurate to say that the US intention was self-interested, but there was some additional moral motivation and appeal of the humanitarian veneer that might be put on the intervention.

26 It could be argued that the ICC is not acting from bias, as these cases were sometimes referred by other Africans. Nevertheless, the lesson remains that one-sided interventions create a perception of bias that undermines the institution.

27 One might object that this discussion hinges on the issue of legitimate authority (which I discuss in Chapter Eight below) rather than just cause. I would reply that the typical legitimacy of the pursuit of a kind of cause has an effect on whether that cause should be considered just, such that there is an interplay between the principles. Questions about humanitarian intervention's legitimacy recommend a strict just cause threshold.

28 My modified legalism, with intervention for exceptional cases, is influenced by Walzer as well as the R2P. Walzer cites the old-fashioned legal standard which allowed intervention to stop "crimes that shock the moral conscience of mankind" (1977, 107). This emotional test is too relative, though it is on the right track in limiting intervention to extreme cases. The list of crimes given by the R2P

helpfully expands on Walzer's short list of genocide and slavery (International Commission on Intervention and State Sovereignty 2001).

29 In Chapter Nine, I address in detail the practical implications of my view for human rights crises today, including those in Syria, the Congo, Yemen, and South Sudan.

References

Ayoob, Muhammad. 2002. "Humanitarian Intervention and State Sovereignty." *The International Journal of Human Rights* 6 (1): 81–102. 10.1080/714003751

Banerjee, Abhijit, and Esther Duflo. 2019. *Good Economics for Hard Times: a Survey of Proven Solutions*. New York: Public Affairs.

Barnett, Thomas. 2004. *The Pentagon's New Map: war and Peace in the Twenty-First Century*. New York: The Berkley Publishing Group.

Barnett, Thomas. 2005. *Blueprint for Action: a Future Worth Creating*. New York: Berkley Publishing Group.

Bellamy, Alex. 2014. "The Responsibility to Protect and the Problem of Regime Change." In *The Ethics of Armed Humanitarian Intervention*, edited by Don Scheid, 166–86. Cambridge: Cambridge University Press.

Bellamy, Alex, and Edward Luck. 2018. *The Responsibility to Protect: From Promise to Practice*. Cambridge: Polity Press.

Bellamy, Alex, and Paul Williams. 2010. *Understanding Peacekeeping*, 2nd edition. Cambridge: Polity Press.

Bellamy, Alex, and Paul Williams. 2012. "On the Limits of Moral Hazard: the 'Responsibility to Protect', Armed Conflict and Mass Atrocities." *European Journal of International Relations* 18: 539–71.

"Between Africa and ICC [editorial]." *allAfrica.com*, November 10, 2016. *Gale In Context: Global Issues*. Accessed October 22, 2020. https://link.gale.com/apps/doc/A469648195/GIC?u=tel_a_apsu&sid=GIC&xid=0253f6c7

Bohman, James. 2008. "War and Democracy." In *War: Essays in Political Philosophy*, edited by Larry May, 105–23. Cambridge: Cambridge University Press.

Brandt Ford, Shannon. 2013. "*Jus Ad Vim* and the Just Use of Lethal Force Short of War." In *Routledge Handbook of Ethics and War*, edited by Fritz Alhoff, Nicholas Evans, and Adam Henschke. New York: Routledge.

Buchanan, Allen. 2001. "From Nuremberg to Kosovo: the Morality of Illegal International Legal Reform." *Ethics* 111 (4): 673–705.

Buchanan, Allen. 2007. "Justifying Preventive War." In *Preemption: Military Action and Moral Justification*, edited by Henry Shue and David Rodin, 126–42. Oxford: Oxford University Press.

Buchanan, Allen. 2010. "Institutionalizing the Just War." In *Human Rights, Legitimacy and the Use of Force*, edited by Allen Buchanan. Oxford: Oxford University Press.

Bull, Hedley. 1977. *The Anarchical Society: The Study of Order in World Politics*. New York: Columbia University Press.

Burke, Anthony. 2005. "Against the New Internationalism." *Ethics & International Affairs* 19 (2): 73–89.

Cassese, Antonio. 2005. *International Law*, 2nd edition. Oxford: Oxford University Press.

Coady, C.A.J. 2003. "War for Humanity: a Critique." In *Ethics and Foreign Intervention*, edited by Deen Chatterjee and Don Scheid, 274–95. Cambridge: Cambridge University Press.

Deaton, Angus. 2013. *The Great Escape: Health, Wealth and the Origins of Inequality*. Princeton University Press.

Diamond, Jared. 1997. *Guns, Germs and Steel*. New York: Norton.

Duffy, Helen. 2015. *The 'War on Terror' and the Framework of International Law*, 2nd edition. Cambridge: Cambridge University Press.

Ellsberg, Daniel. 2003. *Secrets: A Memoir of Vietnam and the Pentagon Papers*. New York: Random House.

Fiala, Andrew. 2008. *The Just War Myth: The Moral Illusions of War*. Lanham, Maryland: Rowman and Littlefield.

Finkelstein, Claire, Jens D. Ohlin, and Andrew Altman, eds. 2012. *Targeted Killings: Law and Morality in an Asymmetrical World*. Oxford: Oxford University Press.

Franck, Thomas. 1988. "Legitimacy in the International System." *American Journal of International Law* 82 (4): 705–59.

Friedman, Thomas. 1999. *The Lexus and the Olive Tree: Understanding Globalization*. New York: Farrar, Straus, and Giroux.

Friedman, Thomas. 2003. "Because We Could." *New York Times*, June 4, 2003. https://www.nytimes.com/2003/06/04/opinion/because-we-could.html

Friedman, Thomas. 2005. *The World Is Flat*. New York: Farrar, Straus, and Giroux.

Gallucci, Robert. 2003. *Lecture on just war and diplomacy, National Endowment for the Humanities Institute on "War and Morality in the Twenty-First Century,"* U.S. Naval Academy, June 2003.

Hagan, Kenneth, and Ian Bickerton. 2008. *Unintended Consequences: The United States at War*. London: Reaktion Books.

Hartmann, Christof. 2017. "ECOWAS and the restoration of democracy in the Gambia." *Africa Spectrum*, vol. 52 (1): 85+. *Gale In Context: Global Issues* Accessed June 5, 2020. https://link-gale-com.ezproxy.lib.apsu.edu/apps/doc/A4994 94915/GIC?u=tel_a_apsu&sid=GIC&xid=b719ed74.

Hathaway, Oona, and Scott Shapiro. 2014. "Outcasting: Enforcement in Domestic and International Law." In *Philosophy of Law,* 9th edition, edited by Joel Feinberg, Jules Coleman, and Christopher Kutz, 299–314. Boston: Wadsworth.

Hedges, Chris. 2002. *War is a Force that Gives Us Meaning*. New York: Anchor Books.

"Helmuth von Moltke." 2012. In *Oxford Essential Quotations*, edited by Susan Ratcliffe. Oxford University Press, https://www.oxfordreference.com/view/10.1 093/acref/9780191826719.001.0001/q-oro-ed4-00007547.

Hoag, Robert. 2007. "Violent Civil Disobedience: Defending Human Rights, Rethinking Just War." In *Rethinking the Just War Tradition*, edited by Michael Brough, John Lango, and Harry van der Linden. Albany: SUNY Press.

Holzgrefe, J.L. and Keohane, Robert, eds. 2003. *Humanitarian Intervention: Ethical, Legal, and Political Dilemmas*. Cambridge: Cambridge University Press.

Hudson, Kimberly. 2009. *Justice, Intervention, and Force in International Relations: Reassessing Just War Theory in the 21st Century*. London and New York: Routledge.

International Commission on Intervention and State Sovereignty. 2001. *The Responsibility to Protect*. Ottawa: International Development Research Centre.

Johnson, Chalmers. 2004. *Blowback: The Costs and Consequences of American Empire*. New York: Holt.

Johnson, Dominic. 2004. *Overconfidence and War: The Havoc and Glory of Positive Illusions*. Cambridge, MA.: Harvard University Press.

Kahneman, Daniel. 2011. *Thinking Fast and Thinking Slow*. New York: Farrar, Straus, and Giroux.

Krugman, Paul. 2018. *International Economics*, 11th edition. New York: Pearson.

Kuperman, Alan J. 2008. "The Moral Hazard of Humanitarian Intervention: Lessons from the Balkans." *International Studies Quarterly* 52: 49–80.

Lango, John. 2014. *The Ethics of Armed Conflict*. Edinburgh: Edinburgh University Press.

Luban, David. 2002. "Intervention and Civilization: Some Unhappy Lessons of the Kosovo War." In *Global Justice and Transnational Politics*, edited by Pablo De Greiff and Ciaran Cronin, 79–115. Cambridge, MA: MIT Press.

Luban, David. 2008. "War Crimes and the Law of Hell." In *War: Essays in Political Philosophy*, edited by Larry May, 266–88. Cambridge: Cambridge University Press.

McMahan, Jeff. 2010. "The Just Distribution of Harms Between Combatants and Non-combatants." *Philosophy & Public Affairs* 38: 342–79.

Morgenthau, Hans. 1954. *Politics Among Nations: The Struggle for Power and Peace*, 2nd edition. New York: Knopf.

Nardin, Terry. 2005. "Humanitarian Imperialism: Response to 'Ending Tyranny in Iraq'." *Ethics & International Affairs* 19 (2): 21–6.

New York Times editorial. 2019. "The Secret Death Toll of America's Drones," March 30, 2019, https://www.nytimes.com/2019/03/30/opinion/drones-civilian-casulaties-trump-obama.html.

Niebuhr, Reinhold. 2008. *The Irony of American History*. Chicago: University of Chicago Press.

Orend, Brian. 2002. "Justice After War." *International Affairs* 16 (1): 43–56.

Orend, Brian. 2013. *The Morality of War*, 2nd edition. Buffalo, NY: Broadview Press.

Orend, Brian. 2014. "Post Intervention." In *The Ethics of Armed Humanitarian Intervention*, edited by Don Scheid, 224–42. Cambridge: Cambridge University Press.

Øverland, Gerhard. 2011. "High Fliers: Who Should Bear the Risk of Humanitarian Intervention?" In *New Wars and New Soldiers: Ethical Challenges in the Modern Military*, edited by Paolo Tripodi and Jessica Wolfendale, 69–86. Farnham: Ashgate.

Pattison, James. 2014. "Bombing the Beneficiaries: the Distribution of the Costs of the Responsibility to Protect and Humanitarian Intervention." In *The Ethics of Armed Humanitarian Intervention*, edited by Don Scheid, 113–30. Cambridge: Cambridge University Press.

Peterson, Richard. 2006. "Human Rights and the Politics of Neo-Colonial Intervention." *Philosophy Against Empire: Radical Philosophy Today* vol. 4, edited by Tony Smith and Harry van der Linden, 211–34. Philosophy Documentation Center.

Piketty, Thomas, translated by Arthur Goldhammer. 2014. *Capital in the 21st Century*. Cambridge, MA: Harvard Press.

Pirnie, Bruce, and Edward O'Connell. 2008. *Counterinsurgency in Iraq*. Santa Monica, CA: Rand.

Recchia, Stefano. 2017. "Authorising Humantiarian Intervention." *Review of International Studies* 43 (1): 50–72.

Regan, Patrick. 2000. *Civil Wars and Foreign Powers*. Ann Arbor: The University of Michigan Press.

Ricks, Thomas. 2007. *Fiasco: The American Military Adventure in Iraq*. New York: Penguin Books.

Rocheleau, Jordy. 2008. "Ethical Principles for State Building." In *Stability Operations and State-Building*, edited by Greg Kaufman, 18–32. Carlisle, PA: Strategic Studies Institute.

Rodin, David. 2014. "Rethinking Responsibility to Protect." In *The Ethics of Armed Humanitarian Intervention*, edited by Don Scheid, 243–60. Cambridge: Cambridge University Press.

Seybolt, Taylor. 2007. *Humanitarian Military Intervention: The Conditions for Success and Failure*. Oxford: Oxford University Press.

Sinnott-Armstrong, Walter. 2007. "Preventive War – What is it Good For?" In *Preemption: Military Action and Moral Justification*, edited by Henry Shue and David Rodin, 202–21. Oxford: Oxford University Press.

Stiglitz, Joseph, and Linda Bilmes. 2008. *The Three Trillion Dollar War*. New York: Norton.

Teson, Fernando. 1988. *Humanitarian Intervention: An Inquiry into Law and Morality*. Dobbs Ferry, NY: Transnational Publishers.

Teson, Fernando. 2005. "Ending Tyranny in Iraq." *Ethics & International Affairs* 19 (2):1–20.

Teson, Fernando, and Bas Van der Vossen. 2017. *Debating Humanitarian Intervention*. Oxford and New York: Oxford University Press.

United Nations. "Preventing Sexual Exploitation and Abuse." Accessed October 24, 2020. https://www.un.org/preventing-sexual-exploitation-and-abuse/content/news-articles.

Valentino, Benjamin. 2011. "The True Costs of Humanitarian Intervention: the Hard Truth about a Noble Notion." *Foreign Affairs* 90 (6): 60–73.

Walzer, Michael. 1977. *Just and Unjust Wars*. New York, NY: Basic Books.

Werner, Richard. 2013. "Just War Theory: Going to War and Collective Self-Deception." In *Routledge Handbook of Ethics and War*, edited by Fritz Allhoff, Nicholas Evans, and Adam Henschke, 35–46. New York and London: Routledge.

Whitlock, Craig. 2019. "The Afghanistan Papers: at War with the Truth," *The Washington Post*, Accessed December 9, 2019. https://www.washingtonpost.com/graphics/2019/investigations/afghanistan-papers/afghanistan-war-confidential-documents/

6 Why two wrongs cannot make a right use of force: a critique of compound just causes

6.1 Introduction

Wars are frequently fought for multiple purposes. In some cases, each of these aims would independently constitute a just cause for a war. For example, either defending against Axis aggression or stopping the Holocaust could have served as a just cause for the Allied fight in World War II. In other cases, only one war aim may be a just cause, such that questions arise whether other aims may be simultaneously pursued militarily. For example, a state may invade a repelled aggressor for national defense while also aiming to accomplish additional punitive or rehabilitative ends.[1] There is also a third sort of case of proposed just cause in which *none* of the various rationales for waging war could *independently* serve as just cause, but it is proposed that they do so *in combination*. A war might be justly fought to respond simultaneously to several injustices, though it would be unjust to take up arms in response to any of them singly. "Compound causes"[2] of this sort are commonly given to shore up the political justification of war when there is no clear single justifying cause. One can also find compound justifications for war by ethicists, and at least a couple of just war theorists have recently explicitly defended compound causes. However, there has been surprisingly little explicit discussion in the just war literature, such that this problematic issue remains undertheorized. I will argue that causes cannot generally combine to justify wars in the manner assumed and articulate limited conditions for compound just causes.

Recent major conflicts give prime examples of the appeal to compound justifications. The United States famously cited numerous justifications for the 2003 invasion of Iraq. Among other aims, these prominently included preventing the Hussein regime's possible future development and use of weapons of mass destruction, punishing Iraq's past crimes (including aggression against Kuwait, chemical weapons use, and attempted assassination of former President George H.W. Bush), and replacing the repressive, rights-violating Hussein regime with a liberal, democratic state. Such compound justifications were offered not just by the George W. Bush Administration but also by intellectual defenders of the Iraq War. Thomas Nichols

DOI: 10.4324/9781003105381-6

summarized his justification in *Ethics & International Affairs* at the time of the invasion as follows:

> The record provides ample evidence of the justice of a war against Saddam Hussein's regime. Iraq has shown itself to be a serial aggressor led by a dictator willing to run imprudent risks, including an attack on the civilians of a noncombatant nation during the Persian Gulf War; a supreme enemy of human rights that has already used weapons of mass destruction against civilians; a consistent violator of both UN resolutions and the terms of the 1991 case-fire treaty, to say nothing of the laws of armed conflict; a terrorist entity that has attempted to reach beyond its own borders to support and engage in illegal activities that have included the attempted assassination of a former U.S. President; and most important, a state that has relentlessly sought nuclear arms against all international demands that it case such efforts (2003, 25).

Nichols' argument is questionable in part because of its emphasis upon punishing already completed injustices and preventing others feared in the future, causes which I criticize in other chapters of this text.[3] However, Nichols' case is also noteworthy in describing just cause as resulting from a long list of grievances, which are proposed as sufficient in compound though not separately.

Another defender of the Iraq War, David Mellow, describes it as a humanitarian intervention justified by the goal of protecting human rights. His proposed just cause is nonetheless a compound of two different sets of human rights violations: on one hand, the Hussein regime's violation of the collective rights to self-determination of oppressed groups including the Shi'a, Kurds, and Marsh Arabs, and on the other, individual basic human rights violations in forms such as arbitrary arrest, murder, and torture. Mellow concludes that "there were both collective and individual considerations that when combined, and probably even when considered separately, were of sufficient moral importance to justify third party military intervention" (2006, 298). This argument shares with Nichols' the view that though there may have been no single wrongdoing, the combat of which justified the invasion, just cause was realized by the combination of the injustices to be combatted.

The long US intervention in Afghanistan also displayed at least two different goals. The initial US intervention after September 2001 was primarily self-defensive, aimed at stopping Afghanistan from being used as an Al Qaeda base for continuing terror attacks. However, the US intervention was also described as a humanitarian intervention that would liberate Afghanistan from the Taliban's repressive governance, especially its violations of women's rights. As Al Qaeda dispersed and time passed without further terror attacks launched against the US from Afghanistan, defense became increasingly questionable as a justification for ongoing intervention there. The continued occupation of Afghanistan probably would not have been justifiable without the added humanitarian aim. However, it is doubtful whether democratization and advancing women's equality alone could justify either the initial invasion or ongoing intervention. Hence, a compound of the two causes was offered by the Obama administration to justify the continuing war. In a 2009 speech, Obama defended a just cause in Afghanistan based on a primary aim "to disrupt, dismantle and defeat al Qaeda in Pakistan and Afghanistan" in order to "protect the American people," while adding a secondary goal of preventing "a return to Taliban rule" which would result in "the denial of basic human rights to the Afghan people – especially women and girls" (Lango 2010, 14). Just war theorist John Lango defends the possibility of justifying ongoing US-Afghan intervention based on a compound of these two causes: although each Afghan war aim individually may

> not [be] by itself sufficiently clear and serious ... the threatened harm to US security of terrorism in combination with the threatened harm to Afghan security of an insurgency in Afghanistan might still be of a kind – and in combination sufficiently clear and serious – to justify *prima facie* the use of military force by the United States. (2014, 17)[4]

Compound causes are also relevant when contemplating future military operations. As I write this at the beginning of 2020, there is a question of whether to have a continuing (or, as some would recommend, increased) US force in Syria. Such an intervention might serve a humanitarian goal of protecting locals from attacks by Isis, the Assad regime, and Turkey. At the same time, the US and allied military mission in Syria could be seen as national defense, combating continued international terrorism, such as the 2015 Paris attacks and other actions coordinated by Isis. Even if one thought that neither of these injustices independently was of sufficient magnitude to justify armed intervention, one might think that they do so in combination.

Common sense and simple math support the idea that multiple insufficient causes could add up to a sufficient cause. Clearly, from a utilitarian or realist standpoint, additional goods to be accomplished provide greater justificatory force. Even from a deontological just war standpoint, there seems to be reason to think that various aims could combine to meet the just cause threshold. After all, even undisputed just causes, like stopping genocide or defending against immediate aggression, could be understood to be composed of smaller aims, such as preventing the murders of individuals. Just causes are, in this sense, *always* compound injustices.

Lango, Mellow, and Nichols are not alone among just war theorists in accepting compound causes. Nicholas Fotion explicitly defends the satisfaction of just cause by a multiplicity of considerations: "several small reasons can, in theory, rise to the level of an overriding reason" (2007, 73), such that "several reasons, each of which is not sufficient to count as a just cause, can be cited as sufficient just cause for war" (113). Other writers, though not specifically defending compound causes, define just cause in a way that permits it to be met by cumulative proportional benefits (Steinhoff 2007).[5] While it has received minimal explicit discussion and is by no means settled doctrine, I take it that acceptance of compound causes is not uncommon and is increasing along with the move from the legalist to the punitive or human rights paradigms as discussed in earlier chapters. That compound causes are frequently invoked in political justification but undertheorized in just war theory make this a pressing issue.

I will argue that distinct war causes cannot usually aggregate into a compound cause with greater justificatory force than the causes have separately. In my view, most invocations of compound causes to justify wars, including those for the invasion of Iraq and the ongoing intervention in Afghanistan, are invalid. I support this by an analysis of the logic of just cause, domestic justice analogies, the implausible implications of compound causes, and the dangerous consequences of their general acceptance.

6.2 The concept of just cause and liability to attack

My principal argument against compound causes stems from what I take to be a plausible and widely accepted analysis of the just cause principle. I only briefly explain and defend this conception, which I assume as relatively uncontroversial for the sake of my argument against compound causes. I take as a reasonable definition of just cause for war that provided by Larry May: "preventing or stopping a wrong committed by state or statelike entity … which is sufficiently morally serious to be analogous to the risk of large loss of life that war involves" (2008a, 57). This entails several sub-criteria. To unpack the concept further, there is just cause for war if and only if:

1 The war responds to an injustice (or set of injustices).[6]
2 The party to be attacked bears responsibility for the injustice, making it liable to attack. (See also McMahan 2005).
3 The injustice involves sufficiently large threatened or ongoing human rights violations to justify the extreme coercion and violence of war. (See also Lee 2012, 130; Lango 2014, 107–33).[7,8]
4 The war to be waged aims to stop that injustice. (See also Lango 2014, 108–9).[9]

These criteria clearly rule out wars that are thought to be in the interest of a war wager or are thought to have global utilitarian benefits, but which (1) do not respond to any injustice, (2) target a party that has not made itself liable to attack by perpetrating the injustice,[10] (3) respond to an injustice of insufficient magnitude to justify war, or (4) despite a sufficiently large injustice occurring, are not aimed at stopping that injustice.

Although there is general agreement among just war theorists about these requirements, there is disagreement about how to specify the content of each. This is particularly true of criterion 3 regarding the magnitude of injustice required to justify war. Resolving this dispute was the focus of Chapters Four and Five and will be set aside for now. I will assume a moderate response to this question in which large-scale violations of basic human rights is the threshold required to justify war, with this typically being met either by international aggression or the atrocity crimes of genocide, war crimes, ethnic cleansing, and crimes against humanity (United Nations General Assembly 2005, par 139). Whether a combination of smaller scale injustices can meet this threshold as well is the focus of this chapter.

I take it that this conception of just cause limits the way in which insufficient and hence unjust causes can compound to become just causes. To have a compound just cause, it would have to be that several proposed causes of war individually fail to meet at least one of the criteria above but satisfy them all in conjunction. That is, an attacker would have to have a plurality of war purposes which in conjunction aim to stop a large-scale violation of human rights which (in total) suffices to make the perpetrator liable to attack and justifies the large scale use of force. My contention is that multiple proposed causes, though each of potential value, cannot generally add up to a compound just cause if none of them is independently a just cause.[11] Let us consider the various ways that sets of causes might be independently insufficient to justify war to see the conditions under which they can form a compound cause which is sufficient.

Most straightforwardly, no number of war causes which all fail to satisfy any of the criteria 1, 2, or 4 can satisfy that criterion in compound and thereby amount to a just cause. Injustice, responsibility, and intent are not scalar and thus cannot be achieved through aggregation. To start with the first criterion, if each of two or more situations does not involve any unjust threat (violating criterion 1), then their combination does not involve an unjust threat either. For example, a war might be thought to have an array of potential benefits, such as opening up economic opportunity, improving life in the target state, stimulating the economy of the attacker through military spending, and testing military strategies. No combination of expected benefits amounts to a just cause when an injustice is lacking.

The same holds if there are injustices but the party to be attacked is not responsible for them (failing to satisfy criterion 2). A blameless party is not liable to attack even if its attack would result in the alleviation of injustices of greater magnitude than the aggression itself is likely to perpetrate. For example, if the invasion of an innocent target state would secure resources or satisfy bloodlust, which helped to end a humanitarian crisis elsewhere, it would nonetheless be unjust. The addition of further causes for which the target state carries no responsibility cannot compound to make it liable to attack.

Similarly, no number of injustices perpetrated by an attacked party justifies a war that does not attempt to put an end to those injustices (failing to satisfy criterion 4). If a party attacks an aggressor or human rights abusing state but does not attempt thereby to stop its aggression or human rights abuses, the injustices do not provide a just cause for *this* war. The crimes are objectively present and might provide just cause for a different war that did aim to stop those wrongdoings. For example, one possible retort to Mellow's case that various forms of oppression conducted by the Hussein regime provided a just cause for the Iraq War is that the US invasion did not appear to be planned to protect the rights of the Iraqi people. In this chapter, my emphasis is not to question particular intentions but to say that if a set of causes are not all intended, they cannot compound to form a just cause.[12]

A second class of compound causes has more intuitive plausibility. In this group, each cause falls short of constituting a just cause in different ways, such that it is proposed that in combination, they can satisfy all elements of just cause. So there might be (a) great benefits that could be achieved by war against Y though they involve no injustice, (b) injustices which could be combatted by attacking party Y even though Y is not responsible for these injustices, *and* (c) injustices for which Y is responsible but which attacker X does not aim to prevent in its attack on Y. Such a war would attack a wrongdoer and would alleviate injustices and be expected to achieve benefits. However, in this case, the fragmentation of the responsibility of the target and intent of the attacker leave the war without a just cause. For its attack on Y to be justified, X must intend to stop an injustice by attacking Y. That X has a goal of alleviating injustices other than those perpetrated by Y does not make Y liable to X's attack. It is not enough for it to be the case that Y has perpetrated an injustice that could make it liable to attack. It must be attacked for the reason of that injustice. For example, if a tyrant is perpetrating great injustices and an invader attacks to secure the tyrant's resources in order to use them to combat injustice in a distant part of the world, this attack lacks just cause, even if it yields a net gain in human rights. Not only are the injustice, responsibility, and intent elements of just cause all or nothing, but they also are all necessarily *unitary*, the same injustice must be the responsibility of the target state or group and its cessation the aim of the attacking party. The "injustice" in each of the four just cause sub-criteria must have the same reference to avoid equivocation in the war's justification.

The most plausible cases for a compound just cause are those in which criterion 3, sufficient magnitude of injustice to warrant war, is not met by any of the individual causes but would be met by them in combination. Magnitude, unlike injustice, responsibility, and intention, is scalar in nature and can aggregate. As I suggested earlier, it is in this sense that there are clearly compound just causes, as any injustice which warrants war will be made up of smaller injustices. A genocide, the paradigm of a just cause for war, is composed of individual murders which would not singly be a cause for war but become one when compounded. Aggression, justifying defensive war, is not created by a single shot or soldier crossing the border but rather by a sufficient quantity of infringement of territorial integrity collectively constituted by many individual actions. Every just war responds to a large injustice which is composed of a compound of smaller injustices in this manner.[13]

However, small injustices cannot always add up to a just cause. To begin with the obvious, some sets of injustices in combination still will not be of sufficient magnitude to justify war even if we accept their aggregation in principle. The set of grievances that many states have against their neighbors would not be sufficient to justify war even in compound.

In other cases, there may be various causes that, in combination, are of sufficient magnitude to justify war but fail to amount to a just cause because the other three criteria do not apply uniformly. If there are injustices which combined are of sufficient magnitude, and putting an end to all of them is intended by a war-wager, it may nonetheless be the case that the party to be attacked is not responsible for all of these injustices, in which case it lacks liability to attack. For example, say State X commits some human rights violations insufficient to justify war, and X is also used, against its will, as safe haven for terrorists, such that a war to overthrow the government of X could end both the internal rights violations and terrorism. The compound of rights violations to be stopped might be of sufficient magnitude to justify war, but the fact that X is not responsible for the terrorism means that it is not liable for a crime sufficient to justify war against it.[14]

A war will also lack a just cause if there are several injustices which, in combination, are of sufficient magnitude to justify war, but the attacker does not intend to right this whole array of injustices.[15] If the attacker has no intention of stopping some of the responsible party's crimes, these actions cannot be part of the justifying cause for *that* war. Thus they cannot contribute to the satisfaction of the magnitude requirement.[16] If a set of human rights violations is necessary to justify an intervention, but the intervention makes no attempt to stop these violations, it is unjustified. If an intervener sought only to put a stop to economic corruption while failing to address any of the other human rights violations which justified its war, its intervention would not have a sufficient cause. To refer back to one of our initial motivating cases, if one thought that the sum of ongoing Iraqi human rights violations listed by Mellow were all necessary and collectively sufficient to justify war, the Bush Administration's Iraq War would still lack just cause if it did not intend to put an end to *all* of these violations. The sort of failed compound cause we are talking about here violates criteria 3 and 4 in combination. The actually intended war would lack sufficient magnitude of injustice being combatted, violating criterion 3. Alternatively, one could also say that the justifying cause of sufficient magnitude is not fully intended, violating 4.

For a war to stop a compound of injustices to have a just cause, it must be the case that a single party is responsible for the full magnitude of causes and that the intended war seeks to put a stop to all of them, such that sub-criteria 1 through 4 all apply in concert. Lest this seem like an impossible confluence, it was quite clearly met by the Allied forces in World War II. The Allies targeted the Axis forces, who were liable to attack through responsibility for large-scale injustices of aggression and human rights violations of sufficient magnitude to justify war, and ending these injustices was the aim of the Allied war. Thus criteria 1 through 4 all applied uniformly with regard to either a war of stopping Axis aggression or one of ending Axis human rights violations.

6.3 The limits of compounding magnitudes of injustice for the justification of war

What then of cases in which the same party commits injustices of sufficient combined magnitude, and war is planned to attack that party to stop all of those injustices? This is the most defensible form of compound cause and is the sense in which every just cause is a compound, including the World War II causes just mentioned. Arguably, it is also the sort present in the Iraq and Afghanistan cases. However, there are ways in which injustices can fail to combine to create a compound cause for war. This occurs when the injustices committed by a party are unrelated such that they do not combine to form one large-scale injustice of the magnitude that justifies war but remain separate insufficient injustices. The party is responsible for each of its threatened injustices and is liable to the sum of the sanctions warranted by each of those crimes. However, it is not necessarily responsible for an overarching crime of simultaneously committing all of these different rights violations, and thus of magnitude needed to justify war.

Above, I left it open whether the injustice in criterion 1 was singular or a set of combined injustices. However, I now argue that the injustice must be understood as a single overarching crime meeting the just cause threshold in order to justify war. Typically, it is understood that only aggression against another state or an atrocity-level violation of human rights is just cause for war. Thus, to justify war, a series of injustices must intelligibly compound to constitute atrocity or aggression. In addition to being necessary for adding up to liability, the requirement that the war simultaneously aims to achieve the justifying cause (criterion 4) implies that we have to be able to think of the cause as pursuable as a coherent whole. To do this requires that the wrongdoings are appropriately related such that they fall under an overarching injustice whose combat is the aim of the war.

My view contrasts with Fotion's case for compound or, as he calls them, "multiple" causes. He notes that his position modifies the traditional paradigm requiring meeting a high threshold of defense against aggression or humanitarian catastrophe (2007, 113). In his detailed example of an intuitively valid compound cause, Fotion presents a collection of actions, including shots fired across a border, the sinking of a ship, the hacking of computers, and terrorist actions against individuals living abroad as arguably constituting, in compound, a just cause for war (76). My interpretation of this is that at this point, the series of actions is seen as constituting an act of aggression which provides the cause for a war of defense, not the list of individual actions. Fotion has not given an example of a compound just cause in which actions justify war without being part of a single overarching recognized crime. If multiple events are unrelated and none meet the threshold, then there is no just cause. If the events are related and satisfy the threshold in compound, then just cause is met but is not essentially a compound (or "multiple" in Fotion's terminology) cause.

An analysis is needed of the sorts of circumstances that can combine to create a just cause, as I have acknowledged they sometimes do. My proposal is that the actions only combine to create a justifying cause when they are part of a single overarching injustice. This usually will mean that the actions are different parts of a single large criminal *plan* or *policy*, such as genocide or aggression directed by a government or group leaders—the various individual murders or acts of ethnic cleansing compound to form the overarching atrocity crime. A compound cause might also exist without an overarching plan if many human rights abuses are interrelated such that they are similar phenomena resulting from the same policy. A government that has a policy of torturing and killing opponents can be seen as engaged in a large-scale injustice even if its killings are not part of a genocidal conspiracy to eradicate a particular group or achieve a particular political purpose. The question arises whether a large number of uncoordinated individual murders could amount to an atrocity, and thus a compound just cause, without an overarching plan or policy. Genocides like those in Armenia, Rwanda, and Darfur, were carried out in part by disparate private individuals rather than a single state operation.[17] In response, I take it that genocidal or other atrocity level violence has generally had organizers and planners, and it is part of the definition of genocide, in particular, that there is a plan.[18] However, I want to allow for the possibility for disparate actions to compound into an atrocity crime. Mob violence, committed by various individuals without a formal policy or leader, could involve similar and mutually influencing cascading rights violations, such that they are an overarching phenomenon and properly described as a single atrocity, either a crime against humanity or a war crime. If the crime were large enough, the perpetrators collectively would be liable to have war waged against them.[19]

However, when a party's crimes are unrelated, not part of a single plan, principle, or pattern, they do not form a compound crime graver than the individual crimes. In particular, they do not add up to an overarching crime constituted by the sum of all the parts, which would make the perpetrator liable to attack and justify war against it. For example, say that a government is guilty of significant corruption, occasional excess violence like murders of political opponents, and a failure to provide adequate positive protection for people's material rights, such that the combined unjust harm is roughly equal to the magnitude of a large-scale killing that is agreed to be a just cause for war. I submit that the three do not combine to form one great injustice that could make the government liable to attack. A perpetrator of multiple injustices is liable to the sum of the sanctions warranted by each of its crimes applied separately, but not for a compound higher crime. I will return to discuss examples of this failure, including the Iraq and Afghanistan cases, below.

This analysis of limited aggregation of liability to attack is supported by a consideration of principles and intuitions in two areas: domestic analogies of cases of individual justice, on the one hand, and an assessment of the nature and purpose of just war theory and international norms on the other. I proceed to those now before returning to discuss the implications for compound just causes.

6.4 Domestic analogies and compound causes

Fotion suggests compound just causes for war are in line with our ordinary reasoning. In most complex decisions, including "personal, business, and academic matters," he points out, there is a series of reasons for one's choice (2007, 78). Any one or collection of positive and negative considerations might tip the balance toward one taking a particular action, like picking a school or buying a car. However, I would reply that these decisions are importantly different from a decision that involves infringing the rights of others, especially their right to life or other basic freedoms. To justify killing or coercing another person, it is not enough to foresee benefits or have a list of grievances. Rather, law and morality require that a substantial and clear threshold be reached before force is justified. The threshold requires either that single grievances are tested to see if they meet the bar or, if a compound, that the multiple grievances are relatable such that they can add up to and be measured as a compound grievance.

Fotion specifically cites "getting a divorce" as a decision in which parties would frequently have a long list of considerations, with no single one being decisive. This example seems to be akin to coercion insofar it involves changing a legal status (dissolving the marriage contract) and ending a committed relationship which, morally, as well as legally, there is a presumption of upholding. I would note that for this reason, some jurisdictions have a standard of "fault" in divorce (such as abuse or adultery) or at least require that some recognized cause be given. Fotion would probably counter that in many states, "no fault" divorce can be initiated for reasons such as "incompatibility" or "irreconcilable differences," which do not suggest a single high threshold and invite a series of considerations. I take it that this is because societies have decided that individual autonomy ultimately trumps the marital vow and that the value of allowing individuals to escape an unhappy marriage outweighs the possible harm of undermining marital commitments. By contrast, in the case of violent coercion, the presumptive obligation of non-violence and non-interference is not easily overridden by either the autonomy of a party who would break the peace or by the potential utility in doing so. Various grievances and potential benefits of fighting do not readily compound to justify violence. Fotion defends his compound causes with the thought that "complex" decisions can rightly involve complex causes. However, I would reply that it is not complexity but rather the lack of a decisive factor that points toward compound rationales. In ordinary, simple decisions, many factors can play a role, while decisions involving the infringements of basic rights and other norms require a definite (and high) threshold.[20]

Our understanding of the liability to force in the face of compound unjust threats can be tested in domestic cases of individual violence and punishment. If an individual threatens several unrelated unjust rights violations to different individuals, each of which is short of the threshold of the magnitude of threat required to justify resort to lethal force against that individual, their conjunction does not generally justify lethal force. For example, say that an individual is assaulting and robbing several people – or kicking thousands in the shins – without the threat of lethal force but with total harm in terms of disutility equivalent to that typically caused by murder. I take it that our intuition is that the individual is not liable to lethal force to stop this rampage, although they would be liable to the lesser degrees of force warranted by the individual transgressions. In assessing justified self-defense, we look at what is warranted by a particular injustice, not by an all-things-considered evaluation of an aggressor's life conduct and the likely results of using force. It is true that in typical domestic cases, one does not know an assailant's other crimes such that it occurs to us to take action to stop them all. However, I take it that even if one did have this knowledge, we would not view the distinct injustices to compound to a massive injustice warranting a lethal response.

In criminal law, unrelated crimes are not aggregated into a compound crime for the sake of determining an appropriate punishment. The commission of several misdemeanors, warranting minor punishment, is not understood to combine to equal a felony, even if one might think that the compound of transgressions was as bad, all things considered, as a felony. A criminal can be charged and punished for each minor offense separately – and in this serial sense, the penalties compound – but does not become guilty of a higher level compound crime warranting a more serious form of punishment than the misdemeanors serially. The international law and moral understanding of the justification of war are similar. Only particular types of crimes – specified in international law as aggression and atrocity-level rights violations (analyzed under the Responsibility to Protect as falling under one of four types) warrant the extreme measure of military force.[21]

One may object that crimes are recognized as more serious when they are compounded by "aggravating factors." A criminal who kills in the course of an armed robbery is sentenced more severely than one who commits a non-violent robbery and an unrelated act of manslaughter. My analysis of this is that the aggravating factors are part of the single criminal action and change the nature of the crime. The action of killing recklessly during an armed robbery is of a different and more grave form than either non-violent robbery or manslaughter or their simple addition. Domestic law, to the extent that it is defensible, accords with my framework of treating liability as resulting from the commission of individual crimes rather than a compound of disparate ones.

Admittedly, civil law treats *repeat* offenders more seriously than first-time offenders. A third misdemeanor is sometimes even treated as a felony (Katbi 2014). This seems to be a case of a compounded cause creating increased liability, contrary to my thesis. In response, I would argue that committing a crime repeatedly, especially after having been found guilty and put on probation, may be properly considered a more serious act and thus warrant a more serious penalty. Even if the serial actions are not formulated as a single crime, such as terrorizing a single individual or group, the fact that the perpetrator continues to commit the crime suggests heightened responsibility and thus greater liability for punishment.[22,23] By contrast, the commission of unrelated crimes does not compound to make the perpetrator guilty of a higher level crime. He or she is not liable to a greater punishment than that warranted by the aggregation of the sentences for the two crimes. Of course, one might argue that the act of "committing a crime" is the sort of thing that, if done repeatedly, creates a new level of liability. This is what has been done in the cases such as "three strikes and out" rules, where life prison sentences are automatically given after a third criminal conviction (Adams 2013, 528–34). I take it that these rules are *prima facie* at odds with justice, especially in cases in which several of the crimes are relatively minor, and that such measures are disputed for good reason. At the same time, even if individual crimes were plausibly thought to compound, states have a less unified agency such that it is more problematic to hold them liable for disparate actions. Moreover, as I discuss in the next section, states tend to commit many minor crimes such that ready compounding would be overly permissive in justifying force.

Ordinary moral and legal conceptions of cause for force and punishment support my thesis that unrelated causes, which are not part of a single criminal assault on human rights, cannot compound to create a just cause for war.

6.5 Implications of compound just causes

In addition to analysis of the just cause principle, the case against compounding unrelated crimes to justify war is strengthened by a consideration of the implications of its acceptance. As I argued in earlier chapters, the just cause principle, as the central criterion for *jus ad bellum,* should provide effective restraints on warfare. A view of just cause whose acceptance would lead to the dangerous proliferation of warfare, e.g. Burke and Walzer's (1977, 76) "innumerable fruitless wars," would be invalid even if it were well-based in principle. Thus, I take it that the just cause principle ought to establish a relatively bright-line threshold for the types of situations that are of a kind and magnitude to justify war, providing a bar to the ready resort to force on realist and utilitarian grounds. My "unitarian" conception of just cause is more practical in its restriction of war than accounts in which unrelated injustices created compound just causes for war. The latter would dangerously utilitarianize the just cause principle. Minor injustices might be rapidly compounded to justify the proliferation of belligerence. This would, in effect, lower the bar for going to war, blurring the historically accepted, bright-line threshold of the only defense against aggression or conscience-shocking atrocity crimes justifying war.

Most governments commit various injustices, such that if these are compounded, a case for just cause for war against them could be formulated. If enough minor wrongdoings across time and space and sphere (of moral violation) are taken as a compound, they can easily end up satisfying the magnitude requirement for just cause. David Luban (2004) has pointed out the dangers of justifying preventive war based upon compounded risks. He notes that if there were a small chance of an attack by a hostile power each year, the attack would become probable over the next decades, such that compounding readily justifies preventive strikes. Although Luban's example is specific to the prevention of future injustice, similar implausible results stem from compounding injustices already underway. For example, by most accounts, Iraq's 1990 invasion of Kuwait was unjustifiable aggression. However, many accounts also allow that Iraq may have had valid grievances against Kuwait, including the maintenance of arbitrary boundaries imposed by colonial powers favoring the wealthy Kuwaiti kingdom with disproportionate oil reserves; unforgiving interest rates charged for loans during Iraq's costly war with Iran, from which Kuwait benefited; and slant drilling into Iraqi oil fields (Coates 1997, 151–4). Though none of these injustices is independently sufficient to justify Iraq's war, if their quantities of harm or gravity were combined, they might be equal to an act of aggression. Thus if compound causes were accepted, condemnation, much less the armed overturning, of Iraq's invasion of Kuwait would be difficult to justify. Such a counterintuitive, implausible implication speaks against the compounding of disparate war causes.

One might also catalog a substantial list of injustices committed by the United States: a history of past criminal aggression in places including Vietnam and Iraq; support of coups, dictatorships, and apartheid; *in bello* failures of discrimination and abusive P.O.W. treatment; profiteering from colonialism and arms trade; as well as internal economic, health care, and prison systems which are inhumane and perpetuate inequality, thereby constituting serious injustices. I take it that the injustices on this list are independently insufficient to justify war against the US. However, all combined, the effects of these American injustices may be easily as grave as Iraq's aggression against Kuwait and possibly equal in combined magnitude to an atrocity. Thus, compounding causes imply that there would be just cause to wage war against the US government, either to overthrow it or to otherwise put a stop to these rights violations. But I take it that intuitively there is not a just cause here because of the disparate nature of the grievances.[24] Again, compound causes have counterintuitive if not absurd implications, implying that just cause should be understood as unitary.

One could argue that compound causes do not have to be construed to result in the implications that I just discussed. However, the worry becomes that without relatively clear guidance for how to aggregate, the concept is too vague to provide a plausible or workable conception of just cause. Misapplication of a vague principle is inevitable in the fog of war, as fear of threats, anger at wrongdoing, and the temptation of decisive military action lead potential belligerents to conclude that their cause would be just.[25] Rivals will typically have committed various actual or seeming injustices; if parties are permitted to compound these injustices into a just cause for war, perceived causes for war will multiply. Moreover, if the idea of compound causes were generally accepted, it would invite abuse by parties seeking to justify aggression by listing enough grievances to overwhelm the basic fact that none of these is sufficient cause. Abuse of compounding was done by both the Hussein regime in its aggression against Kuwait and, I shall argue, the Bush administration in its aggression against Iraq. While these justifications were confronted with skepticism on the part of the international community, if the compounding of causes became generally accepted, it would be more difficult to argue against and condemn international aggression.

Implausible implications and dangers of its acceptance tell against the acceptance of the compounding of unrelated causes. My proposal of refusing to compound causes except in the cases of the most clearly related injustices is supported by not only principled analysis but also pragmatism.

6.6 Reply to objections

I anticipate an objection that I am committing a category mistake. Some would argue that I am confusing just cause as an ideal moral principle with what the laws and action guiding rules of war should be.[26] One can imagine that there are principled, just causes that it would nonetheless be dangerous to announce as permitting causes for war due to the potential for misuse, abuse, and destabilization. Compound causes may be of this principled sort. I would suggest two responses. First, I have doubts that principles of justice can be ideally derived and recognized separately from their action guiding functions and the acceptability of their consequences to those affected. Second, even if it makes sense to speak of an ideal theory of justice that holds independently of its consequences, this is not what just war theory, as applied ethics, should be. Just war criteria are needed that can guide decision-making regarding recourse to war as well as after-the-fact assessments of wars' justice for the sake of holding parties accountable. For this purpose, valid just war principles must have acceptable consequences upon application in addition to principled plausibility.[27]

A related objection is that my consequentialist concerns regarding the proliferation of war are relevant to the consideration of the *jus ad bellum proportionality* principle rather than that of just cause. Wise application of the demand that wars only be fought if they would not do harm disproportionate to the cause should prevent much of the misuse of compound causes that I just presented. It might, for example, have ruled out the Iraq War. In reply, I would make several points. First, I find it implausible in principle to say that there is a just cause for military intervention against any state, such as the US, which has multiple injustices on its ledger and that it is only our assessment that such a war might be disproportionate that could rule it out. Second, proportionality judgments are themselves too subjective and imprecise to rely on them as opposed to a bright line, high threshold for limiting war. Although the Iraq invasion was disproportionately harmful in its results, the agents behind the decision expected positive results and may have sincerely – if not wisely – made a proportionality assessment. If we are to avoid an invitation to such military adventures based on expectations of utility, a high bar with relatively bright lines is required. There are utilitarian reasons for refusing to rest the restraint of war on entirely utilitarian grounds. Third, even if we were apt to apply proportionality judgments rigorously, a system in which states widely recognize just cause for war against each other while generally forbearing from following through on proportionality grounds – in addition to being counterintuitive – undermines international cooperation and stability. Like the general acceptance of preventive war, the practice of fighting on compound grounds heightens the likelihood of an unjust attack by one's rivals, thereby giving each incentive for belligerence, leading to a spiral of increasing rationales for war. These dangers give additional reason to conceive of just cause as having a high threshold that cannot be met through a compound of lesser harms or aims.

Surely, one might object, additional moral aims should be able to play some role in contributing to the justification of war. In response, I note, first, that various purposes, though insufficient to count as just causes separately or in combination, could contribute to the *proportionality* of a war that had an independent sufficient just cause. For example, the undermining of the Taliban's repressive control over Afghanistan might have helped to make worthwhile a war whose primary just cause was defense against Al Qaeda aggression.[28] Additionally, if a war is fought for an independently just cause, it may be that in the course of that war or in an occupation during its aftermath, the war fighter is justified or even obligated to use force to prevent other lesser injustices besides the original justifying cause. For example, having overthrown the governments of Afghanistan and Iraq putatively for defensive reasons, the US and allied forces were justified in taking measures to protect the rights of the local population.[29] Furthermore, even if there is no independently sufficient cause and the ongoing injustices do not compound to a just cause – and war would be unjust per the view defended here – it could be justified to take measures short of war (e.g. diplomatic pressure, sanctions, and non-military humanitarian aid) to address these wrongs. As I argued in Chapter Five (Section 5.9), anti-interventionism does not mean doing nothing in the face of injustices until they become full-blown atrocities or aggression.

6.7 Application to Iraq and Afghanistan

To return to the cases that motivated the question of compound causes, I submit that in the case of the Iraq War, the various war causes could not create a compound just cause. The Hussein regime's threatened potential development of weapons of mass destruction, repressive governance, and past injustices, were separate crimes that were not part of a single action on the part of the Hussein government. As I mentioned earlier, whether the intervention was, in fact, intended to combat this whole array of injustices is questionable. However, my point now is that even if it was so intended, the separate injustices do not combine to create a single overarching injustice that can serve as the cause for the war. The war had an array of separate causes, each of which is dubious as just cause. Mellow's two sorts of causes – threats to collective rights and to individual rights – both are human rights violations and thus the sorts of actions that can create just cause for war. However, since these crimes were disparate actions occurring at different times and not part of a single plan or policy on the part of the Hussein regime, they do not aggregate to a single large-scale violation of human rights that could serve as just cause for war. Of course, Mellow and other defenders of the Iraq War (Teson 2005) would reply that the Hussein regime was engaged in systematic disrespect for human rights, meeting the large-scale injustice threshold. However, this is to implausibly lump together many unrelated actions into a single crime. The different actions cannot be described as a single large-scale human rights violation of the sort that justifies war. While my argument in this chapter helps to make the case that the Iraq War was not justified, my thesis is not the injustice of *that* case but rather the establishment of a framework that must be satisfied for any war to be justified. Further discussion of the facts in Iraq and the application of my principle in that light would be needed to judge that case definitively.

The long occupation of Afghanistan has similar just cause difficulties. The intervention sought to both defend[30] against international terrorism and provide armed humanitarian aid to the government in its war against the repressive Taliban. These are two different causes involving the combat of two different injustices. I take it that to a large degree, the perpetrators of the injustices are different: the internally oppressive Taliban fighters were distinct from the Al Qaeda terrorists plotting international attacks. Even if the same core group were substantially responsible for both Afghan repression and international terrorism, the crimes are sufficiently distinct that it doesn't make sense to speak of a single large-scale injustice whose combat justifies war.

In his analysis of the Afghan case, Lango suggests that we can understand these two causes in the Afghan war as compounding into one overarching cause of "countering violent spoilers of peacebuilding" (2010, 18). In this analysis, the fight to restore the Taliban's oppressive governance and the support of Al Qaeda terrorism are violations of basic human rights which combine to form an injustice of sufficient magnitude to justify a war aimed at ending both. First, I take Lango's articulation of an overarching cause to accord with my requirement that causes can only compound if they are intelligible as involving a single purpose combatting a single crime of sufficient magnitude. Secondly, however, I contend that in the Afghan case, the causes were not sufficiently related to create a compound cause. Although the Taliban oppressors and Al Qaeda terrorists are both evildoers antagonizing human rights and peace, I think it is a misnomer to present them as engaged in a single compound injustice. Their rights violations are distinct actions of distinct types, with no overarching plan or underlying principle. Nor are they part of a mass spread of similar and related crimes like the unsupervised mob atrocity that I raised earlier. I take it that to a significant extent, the military responses to these injustices were also distinct: some operations helping Afghan forces win the civil war against the Taliban and others targeting international terrorist bases. In my view, then, the attempt to specify an overarching cause of "countering violent spoilers of peacebuilding" glosses over the differences between the injustices and corresponding war aims rather than exposing an underlying logic of the Afghan mission.[31]

Admittedly, there is an element of interpretation in determining which events are sufficiently related such that they become a compound injustice and thus part of a compound just cause for a war to combat it. Without attempting to offer a precise decision-procedure or a metaphysical account of what constitutes a single action or event, I have suggested criteria that make the determination rational and non-arbitrary. To be part of a large-scale crime that serves as cause for war, a set of injustices must be part of a single plan, policy, or event. However, if there are various injustices that are neither part of an overarching plan or significantly related, then I submit that they should not be conceived of as a compound injustice that is just cause for war. The judgments that, on one hand, Nazi Germany's murders were part of an overarching plan, and on the other that the Hussein regimes' crimes used as a rationale for the 2003 Iraq War were not, are not arbitrary, but result from an assessment of objective facts in terms intelligible categories consistently with ordinary language and moral intuitions. This shows that my conception is not inapplicably vague.

At the same time, I would argue that it is too demanding of a moral concept to expect it to precisely specify its own application. Under that requirement, we would have to reject many other just criteria as well, as concepts such as non-combatancy, proportionality, and last resort have their own ambiguities. A degree of imprecision is typical and does not make a principle empty or inapplicable. Moreover, the idea of compound causes contains an even more basic vagueness about how incommensurable crimes aggregate. In that case, a fundamental vagueness has the dangerous result of permissiveness regarding the recourse to armed force. By retaining a relatively bright-line threshold, I would limit vagueness, and by keeping the threshold high and presumptively not reachable through aggregation, make the remaining vagueness less risky.

6.8 Concluding thoughts on compound causes and the justification of war

In conclusion, I want to compare my view to the position recently defended by Brian Orend, with which it has some similarities. Orend argues that in justifying going to war, a belligerent must "call it's shot," pointing to one justifying cause rather than listing multiple purposes in a "scattershot" manner (2013, 50–2). I share Orend's concern about the too-easy justification of war with a mélange of charges in the hope that some will stick, and I agree with his criticism of the Iraq War in particular on this ground. However, unlike Orend, I accept in principle the possibility of a plurality of justifying causes. For example, I think the Allied entry into World War II had at least two different just causes, and it would have been appropriate for the Allies to cite humanitarian as well as national defense rationales for fighting the Germans and Japanese. What is necessary is that *at least* one of the rationales is sufficient to justify war. My conception of just cause addresses Orend's core Kantian concern, which is that a just cause must provide a coherent rationale for war. The offering of multiple causes without establishing the sufficiency of any of them leaves a war without a defensible foundation. I have argued that this is unjustifiable in principle and dangerous in practice.

If my reasoning is correct, then if the various war aims in Iraq and Afghanistan could not singly justify armed intervention, their combined pursuit through military force was also unjustified. Today, if military intervention is justified, for example, against Isis in Syria, it is because the threshold of sufficient magnitude is met by either a cause of humanitarian intervention or one of international defense but not by a vague amalgam of the two.[32]

Notes

1 Thomas Hurka (2007) controversially defends the pursuit of these other "conditional" just causes alongside the "independent" just cause. For my criticism, see Chapters Three and Seven.

2 I borrow this term for the phenomenon from John Lango (2010, 9–21), who defends such a cause in the case of the ongoing intervention in Afghanistan. The issue is sufficiently undertheorized that there is not yet a widely accepted term of art.

3 See Chapter Three on punitive war and Chapter Nine on preventive war.

4 Lango does not argue that there definitely was just cause, but more minimally maintains that the two purposes could, in principle, be compounded in a way that justifies the war even if neither did independently.

5 Steinhoff writes (25) that "the only logical and practical thing to do is to conceive of the criterion of just cause as sufficient, and hence to let it encompass the criteria of last resort, probability of success and proportionality." I take him to mean that if a group satisfies these criteria, it has just cause, and its war is just. A similar view of just cause as present when a combination of proportionality, last resort, etc. are met is defended by Holliday (2008).

6 The parenthetical qualifier is so as not to beg the question against compound causes. See note 10 below for response to skepticism about the injustice requirement.

7 Some will note that this criterion sounds like the proportionality principle. As I argued in Chapter Four, there is an element of proportionality in just cause, such that there is only a just cause when a war confronts an injustice of a sufficiently large magnitude. Proportionality is not fully entailed by just cause. A particular war might meet this initial magnitude of cause requirement by responding to a sufficiently large injustice, but then be ruled out as disproportionate when its expected nature and effects are taken into account.

8 Some just war theorists would deny that a large-scale injustice is required. I make a case for this in Chapters One, Two, and Four. In this chapter, I assume that a large-scale injustice is required and inquire to what extent this threshold could be met by a compound of lesser injustices.

9 This last requirement introduces a subjective element into just cause, combining just cause with the core of right intention. It can sometimes be helpful to refer to just cause, specifically as the objective injustice to which the war responds, as May does above. This allows us to speak of a war having just cause but lacking right intention. However, there is a sense in which it is a misnomer to speak of a war having an injustice as its justifying cause if it does not intend to stop that injustice. This follows in part from an ambiguity of "cause" as referring to both a grievance giving rise to war and a purpose for which the war is fought. I will sometimes refer to the objective cause as the just cause. However, ultimately I emphasize that the objective and subjective elements both must be present, and linked such that the war aims at stopping the injustice. While there is an element of right intention in just cause, this does not necessarily fully encompass the right intention principle. There may also be elements of motivation or intent that have to be evaluated to determine whether right intention is completely satisfied. My focus in this chapter is an analysis of the justice of compound causes rather than the relationship between the *ad bellum* criteria.

10 Steinhoff (2014) attempts to refute the analysis of just cause as entailing injustice creating liability to attack, by arguing that war could be justified by the necessity to avoid great harm even if there is no wrongdoing making a party liable to attack. However, I would respond that to the extent that such a case does seem to be a just cause for war, it is because the party to be attacked is at least culpably negligent in failing to provide assistance and is committed to defending the

injustice with force. Steinhoff fails to provide an example in which war is justified without responding to injustice.

11 As I acknowledge below, some properly related injustices can compound to make a just cause.

12 For a defense of the necessity of right intention, see Chapter One and note 9 in this chapter.

13 There may be an exception in an extreme case where a party's use of a single weapon of mass destruction, most likely a nuclear bomb, would be a large enough unjust act to serve as just cause on its own.

14 If the state willingly harbored terrorists in addition to committing other human rights violations, it would be responsible for the whole set of crimes. Then, the question would become whether the various crimes combine to form an overarching cause for war. That is the last sort of problematic case that I discuss below.

15 Above I considered the failure of just cause in case the attacking party does not intend the just cause at all. Here, I add that it must intend all the parts of compound just cause whose existence is necessary for the war's justification.

16 This is partly why punitive war is unjustified, as discussed in Chapter Three: although the attacked state is responsible for injustices, the attacking party does not intend to stop these injustices, and the war purpose does not warrant the violence involved in war. See also May (2008a, 59).

17 I am indebted to a reviewer for asking me about my view of such a case.

18 I take it that the Armenian, Rwandan, and Darfur Genocides, although they in part carried out by private forces, were nonetheless in large part directed by government (and perhaps some private) leaders and involved a plan and policy.

19 Whether or not particular individuals can be legally tried and punished for a dispersed atrocity crime is a different matter. For a discussion of the distinction between recognizing injustices and perhaps just cause for war, on one hand, and the justification of criminal prosecution, see May (2008b, 13–7).

20 This discussion does not mean that various injustices aside from the existence of aggression or massive human rights violations are irrelevant to just war theory and international relations. I address this in responding to objections in Section 6.6.

21 On the definition of international crimes, see Duffy (2015, 123–48) and Cassese (2005, 436–50).

22 There are, of course, various theories of just punishment. I take it that a plausible account must contain a deontological element limiting punishment to crimes for which one is responsible, supplemented by consequentialist benefits as well. See Chapter Three and Feinberg, Coleman and Kutz (2014, 689–825).

23 An argument for just cause in the invasion of Iraq could be developed based in part on the past conduct of Iraq regarding the development and use of weapons of mass destruction if there were evidence of its continuing to do so. I take it that more concrete evidence of such a program would have been required, along with a UNSC vote according to international law.

24 There probably was a just cause to defend against the US interventions in Vietnam, Iraq, and elsewhere. However, there is not a just cause to attack the US punitively now, much less try to overthrow the government for these past injustices. Some might argue that there is an overarching policy of US imperialism beneath these past and ongoing injustices. Alternatively, there is a right-wing conspiracy that sees US government actions as a part of a single plan of taking away individual freedom. If either of these is an accurate description of the plans of the US government, the collective actions might be a compound cause for war against the US (contingent on additional criteria such as last resort and

proportionality). However, I take it that these conspiracies are implausible, and the overthrow of the US government is not justified on these grounds.

25 See also, Chapter Five, Section 5.7.

26 Allen Buchanan raises this objection against those who would reject preventive war in principle on the grounds that it would be bad practice to accept it (2007, 130).

27 As I argued in Chapter One, this is not to say that there is no difference between law and morality. An external moral basis for law is required, and there can be moral reasons to break or change laws. Moreover, there may be elements of morality that cannot even be in principle formulated in terms of positive law, such as imperfect duties.

28 On this point, I disagree with McMahan (2005, 79) and Hurka (2007, 199), both of whom have influentially argued that only benefits related to the just cause should count toward meeting the proportionality requirement. In their view, a rationale that could not be justly pursued militarily should not tip proportionality in favor of waging war. In my view, on the other hand, a war must have a sufficient just cause and must also have its expected benefits outweigh its expected harms. Expected benefits unrelated to the just cause could help tip the balance in making the just cause worthwhile to pursue. Of course, decidedly unjust goals should never be pursued by any means, and, to repeat the thesis of this chapter, military means must have an independently valid cause.

29 However, as I argue in Chapter Eight, it is problematic to fight a new war that lacks an independent just cause. That discussion of the ethics of continued war and occupation concludes that to be justified, the additional measures taken in the pursuit of insufficient causes should be actions short of war.

30 There is a question about whether the Afghanistan intervention can be called defensive since the 9/11 terror attacks were completed by the time of the US response. The war thus might be viewed as punitive or preventive rather than defensive. I discuss these sorts of causes in Chapters Three and Nine, respectively, arguing that they are only valid insofar as they can be viewed as part of a defensive aim. In my analysis, the initial US force to disrupt the Taliban and Al Qaeda was arguably justified as a defensive war.

31 I do think it is possible that the different injustices might in some way be interrelated to strengthen the case for a just cause in the Afghan intervention. The threat of the Taliban winning the civil war adds to the terrorist threat. Insofar as the Taliban plan to aid the terrorist threat through coming to power, this gives the US and others a defensive cause in fighting the Taliban that just intervening in Afghan affairs would not have. However, in this case, defense is the justifying cause and not its compound with the humanitarian cause of democratization.

32 In Chapter Nine, I discuss the morality of intervention in Syria and other locations today, arguing that just cause may be met. My argument here has been about the sort of justification that has to be given.

References

Adams, David. 2013. *Philosophical Problems in the Law*, 5th edition. Boston: Wadsworth.

Buchanan, Allen. 2007. "Justifying Preventive War." In *Preemption: Military Action and Moral Justification*, edited by Henry Shue and David Rodin, 126–42. Oxford: Oxford University Press.

Cassese, Antonio. 2005. *International Law*, 2nd edition. Oxford: Oxford University Press, 2005.

Coates, A.J. 1997. *The Ethics of War*. New York: Manchester University Press.

Duffy, Helen. 2015. *The 'War on Terror' and the Framework of International Law*, 2nd edition. Cambridge: Cambridge University Press.

Feinberg, Joel, Jules Coleman, and Christopher Kutz, eds. 2014. *Philosophy of Law*, 9th edition. Boston: Wadsworth.

Fotion, Nicholas. 2007. *War & Ethics: a New Just War Theory*. London: Continuum Books.

Holliday, Ian. 2008. "When is a Cause Just." In *Military Ethics*, edited by C.A.J. Coady and Igor Primoratz, 55–73. Farnham, UK: Ashgate.

Hurka, Thomas. 2007. "Liability and Just Cause." *Ethics & International Affairs* 21 (2): 199–218.

Katbi, Andrew. 2014. "Crossing the Line: An Analysis of Problems with Classifying Recidivist Misdemeanor Offenses as Felonies." *Alaska Law Review* 31 (1): 105–30.

Lango, John. 2010. "Is there a Just Cause for Continuing Military Operations in Afghanistan?" *International Journal of Applied Philosophy* 24 (1): 9–21.

Lango, John. 2014. *The Ethics of Armed Conflict: A Cosmopolitan Just War Theory*. Edinburgh: Edinburgh University Press.

Lee, Steven. 2012. *Ethics and War: An Introduction*. Cambridge: Cambridge University Press.

Luban, David. 2004. "Preventive War." *Philosophy & Public Affairs* 32 (3): 207–48.

May, Larry. 2008a. "The Principle of Just Cause." In *War: Essays in Political Philosophy*, edited by Larry May, 49–66. Cambridge: Cambridge University Press, 2008.

May, Larry. 2008b. *Aggression and Crimes Against Peace*. Cambridge: Cambridge University Press.

McMahan, Jeff. 2005. "Just Cause for War." *Ethics & International Affairs* 19 (3): 1–21. 10.1111/j.1747-7093.2005.tb00551.x

Mellow, David. 2006. "Iraq: a Morally Justified Resort to War." *Journal of Applied Philosophy* 23 (3): 293–310. 10.1111/j.1468-5930.2006.00342.x

Nichols, Thomas. 2003. "Just War, Not Prevention." *Ethics & International Affairs* 17 (1): 25–9. 10.1111/j.1747-7093.2003.tb00415.x

Orend, Brian. 2013. *The Morality of War*, 2nd edition. Buffalo, NY: Broadway Press.

Steinhoff, Uwe. 2007. *On the Ethics of War and Terrorism*. Oxford: Oxford University Press.

Steinhoff, Uwe. 2014. "Just Cause and Right Intention." *Journal of Military Ethics* 13 (1): 32–48. 10.1080/15027570.2014.908647

Teson, Fernando. 2005. "Ending Tyranny in Iraq." *Ethics & International Affairs* 19 (2): 1–20. 10.1111/j.1747-7093.2005.tb00496.x

United Nations General Assembly. 2005 World Summit Outcome Document, paragraphs 138 and 139. Accessed October 24, 2020. https://www.un.org/en/genocideprevention/about-responsibility-to-protect.shtml.

Walzer, Michael. 1977. *Just and Unjust Wars*. New York: Basic Books.

Part III

Just war procedures and application

7 *Jus ad (continuandum) bellum*: reevaluating the justice of interventions over time

7.1 Introduction: just recourse over time

It is commonplace to think of *jus ad bellum*, the justification of recourse to war, as an ethical question to be decided at the time of a party's entry into a war, once and for all. Having decided one is justified in fighting, one's subsequent ethical decisions would then be *jus in bello* questions about appropriate means to adopt in pursuit of one's cause. However, this is to make the problematic assumption that one will continue to be justified in recourse to fighting and killing for this cause at all. New information or changed circumstances can alter the moral status of a war as a whole. Justly initiated armed conflict can become unjust to continue on *ad bellum* grounds. I argue that a belligerent must continually satisfy all the *jus ad bellum* criteria, or it is obligated to stop fighting.

After rejecting the simplistic "one-off" view of just recourse, I dispute a recent strain of theory initiated by Darryl Moellendorf (2008), which frames the question of ongoing just recourse as a matter of "*jus ex bello*," just exit from war.[1] While that approach shares my view that the justification of war needs to be rethought over time, by saying that it is the termination rather than continuation of war that is problematic, it implies a presumption in favor of continued conflict. If we were to adopt a new Latin phrase for decisions about war continuation, we could call it "*jus ad continuandum bellum*,"[2] justice in continuing to be at war. However, since the issue is whether to be at war at all and the conditions are the same as those for the initial just recourse, ultimately I suggest referring to the matter as one of *jus ad bellum*, with the understanding that it is to be continually reapplied. Ethicists, including Brian Orend (2013), Thomas Hurka (2007), and Daniel Statman (2015), along with Moellendorf, defend continuing a war at a lower threshold than it takes to initiate one. Against these views, I maintain that the same criteria apply throughout the course of war, such that it becomes unjust and should be ended if it no longer satisfies any principle of just recourse.[3] I conclude by discussing conditions for justly ending wars, including the terms of peace treaties.

DOI: 10.4324/9781003105381-7

7.2 Continuous vs one-off *ad bellum* justification

Just recourse theory's use of the Latin adjective "ad" in "*jus ad bellum*" connotes a decision about going "to" war, that is, beginning or first entering it. "*In bello,*" by contrast, implies measures undertaken in the course of war. Recently, many have added a third category of *jus post bellum*, justice after war, which includes justice in peace treaties and measures of retribution, reconciliation, and reconstruction. The three moral categories are typically conceived, with Orend, as addressing "three [temporally distinct] phases: beginning, middle and end" (2013, 185). Orend and others sometimes qualify that these moral spheres overlap in various ways. For example, the anticipated manner of fighting may affect the justice of entry into war, and, for revisionists following McMahan (2009), the *in bello* right to kill is contingent on one's *ad bellum* war cause. Nonetheless, there is a commonsense view, reinforced by just war language and supported by much just war theory, that if a party is justified in entering a war, it need not reapply the *jus ad bellum* criteria throughout. This amounts to an assumption that if recourse to war was once justified, with the *ad bellum criteria* satisfied, it will continue to be.

Reflection shows that the "one off" view of *jus ad bellum* is untenable. As I suggested above, "*jus ad bellum*" does not necessarily refer only to justifiably first going to or entering a war. It can and should be understood as referring to having a right to be at war, fighting for one's cause. The just recourse conditions apply not only at war's initiation but also throughout its continuation. A war whose prosecution was once justified can become unjustifiable to continue.

There are at least three different ways that *jus ad bellum* can be lost in the course of a conflict. First, a belligerent may change its war mission, such that it loses its just cause and right intention. For example, as influentially discussed by Walzer (1977, 117–20), in the Korean War, when MacArthur crossed the 38th parallel and began to seek to overthrow the North Korean regime rather than simply defend South Korea against aggression, the US war arguably lost its just cause. Whether or not one agrees with this assessment of the Korean case, one must admit that it is possible for a party that initially fights with a just cause to subsequently adopt an unjust purpose (say collective retribution or domination of the former aggressor or human rights violator). In such cases, it is not the *in bello means* which are problematic but the new war *mission*. The party is no longer justified in pursuing its ends militarily, which is to say it no longer satisfies just recourse (*jus ad bellum*). In these cases of a change in mission, we could speak of there being a new war. However, ordinary language treats a temporally and spatially continuous deployment of military force as a single war. Historians tend to refer to a single Korean War rather than two separate wars, one defensive and one aggressive. In the end, it is arbitrary whether we speak of a just war followed by an unjust one or a just phase followed by an unjust phase of the same war. Either way, *jus ad bellum* status changes over time.

The change of a war cause without adequate reflection is frequently termed "mission creep." It is common in warfare because once a mission is accomplished or there are other changes in circumstances affecting the attainability of goals, a party involved in the war is apt to ask what the best thing to do now is, "while we're here." Mission creep has a bad name largely for its history of getting intervening parties embroiled in endless wars which are ultimately at odds with their own interests. Morally, the new mission must pass not just a realist[4] test (although prudential concerns are relevant to the moral categories of proportionality and likely success) but must meet moral conditions for the use of force. It is easy for realist thinking to dominate any war decisions, and a party that has already determined its war mission to be justified is apt to keep fighting without questioning the justice of its recourse to war. However, it is important that moral, just war questions be raised whenever an evolving mission is contemplated. The pursuit of aims that could not have justified going to war in the first place – from naked national interest and vengeance to prevention and democratization – cannot be a just cause for continuing to fight and kill either.

A change in mission is only one way that a war can become unjust.[5] *Circumstances* can also change so as to make the continuation of a previously just war unjustifiable. Events could make a party cease to meet any of the *jus ad bellum* criteria in the course of a war. In terms of just cause, the most obvious change would involve victoriously achieving the cause, such that there no longer is a large-scale injustice to combat. Short of decisive victory, the justifying cause might be lost in other ways. If an aggressor breaks off its attack or is repelled, then a defensive cause ceases to exist. If human rights abuses are stopped, then an armed humanitarian intervention loses its justifying cause. In such cases, the belligerent party which no longer has a just cause would be obligated to stop fighting.

Perhaps most straightforwardly, if not commonly, *legitimate authorization* might also be lost in the middle of a conflict. In Chapter Eight, I defend the necessity of the legitimate authority criterion. From the standpoint of domestic law and obligations, troops would normally be condemned for carrying on a war after they were ordered to stop by the chain of command. Their continued fight would be criminal. It would be more common for a national force to consider continuing fighting after *international* authorization is withdrawn. If international legitimacy is required for just recourse, as I argue in Chapter Eight, a revocation of authorization by the UNSC or a sanctioning regional body could also make continued war unjust.[6]

One might doubt, with Moellendorf, whether the *last resort* criterion could cease to be met in the course of a war. He argues that since last resort "requires the employment of some alternative means before resorting to war it cannot be the case that a war that once failed to satisfy the principle, could later come to satisfy it" (2008, 127). The same reasoning implies that last resort, once satisfied, could not cease to be satisfied. It is a truism that once war is underway, it can no longer be avoided. However, it is possible for peaceful alternatives to war to arise during the conflict. A diplomatic solution, such as a peace treaty or surrender, may become newly possible.[7] Indeed, such an alternative *must* eventually develop if war is ever to end. If there is an acceptable peaceful solution that realizes the just cause that was the reason for fighting, then a belligerent party is obligated to accept this solution in lieu of continuing to fight. War must (still) be the only viable resort – that is, the last resort – in order to be justly continued. The last resort principle does not need to be interpreted in a one-off manner any more than the other *ad bellum* principles.

In terms of *proportionality*, complicating factors that occur during a conflict, such as developments in strategy, changes in alliance, misfortunes in battle, economic developments, changes in public opinion, or shifts in terrain or weather, may all lead a war to have higher costs than previously foreseen, resulting in it being no longer worthwhile and thus immoral to continue. Casualties among combatants and civilians may turn out to be higher than one projected upon entry into the war. World War I, if it ever was of an import that could justify its fighting by either side, certainly became unjustifiable at some point after it was apparent how costly it would be to achieve victory through the continued trench warfare which came to characterize the conflict.

Similar considerations may also affect the likelihood of success, whether as an independent criterion or as a component of proportionality.[8] A war that initially seemed winnable may subsequently become unwinnable. For example, the entry of China into the Korean conflict made a decisive US victory, even if this were thought to be a just goal, unlikely or at least disproportionately costly in terms of American and Korean lives and welfare to continue. The apparent cost and difficulty of winning ultimately convinced the US to quit the Vietnam War. Even if one thought that the US had just cause and proper authorization to fight that conflict (both of which are dubious), it became clear that the destruction and loss of life the conflict was causing outweighed any likely furthering of moral aims. Regardless of how reasonable it was at the beginning, if and when it is clear that continuing to prosecute war is likely to do more harm than good, one should stop fighting.

In addition to changing circumstances (read broadly to including changing strategies), a change in our evaluation of *jus ad bellum* may arise because of new *knowledge*. It may be that one's war mission already lacked just cause, last resort, likely success, or proportionality, but that one did not realize this. For example, it was likely the case that World War I, Vietnam, and the Second Gulf Wars were not going to be proportionate from the beginning.[9] Once one realizes (or should realize) that one's war lacks a just cause, has viable alternatives, or is not winnable at proportionate costs, one should stop fighting it. This new knowledge could be understood as a variety of changed circumstances.[10] However, since its focus is on epistemic conditions rather than material conditions, it makes sense to distinguish it.

I have emphasized that a previously justified war can become unjustified by ceasing to meet any of the *ad bellum* criteria. The reverse is also the case. An initially unjustified war could become justified with changes in mission, circumstance, or knowledge. I have argued elsewhere (Rocheleau 2010a) that this happened during the 2nd Gulf War, with the US initially invading Iraq unjustly but subsequently being justified in remaining to fight on humanitarian grounds. Since the theme of this text is ways in which war tends to be justified too readily, I focus on cases in which war becomes unjust to continue. Presumably, no party initiates a war that it knows (or even believes) to be unjust, so the question of whether a war can become justified does not usually arise from the standpoint of war wagers. However, a party could realize that its war was initially unjust and inquire whether changed circumstances nonetheless justify continued prosecution. While one should be especially skeptical about rationales offered by a party found to have been fighting unjustly, their validity cannot be dismissed out of hand on principle. The possible favorable reevaluation of a war also can be relevant from the standpoint of an external or historic critic. Discussions below of the continuing justification of war will apply in part to initially unjust wars; I will include considerations of the extent to which initial injustice affects a war's subsequent evaluation.

7.3 Prospective proportionality, sunk costs, and the value of redeeming the dead

In one interesting sense, my conception of continuous *jus ad bellum* could be more permissive in terms of proportionality than a one-off view. Because I argue that a war must be proportionate going forward, my view does not count a war's previous costs against its continuation. Against this, Moellendorf (2015) and Fabre (2015) argue that wars have a proportionality quota established at the beginning, which they may not exceed. For example, if 100,000 casualties were deemed the proportionate cost of an armed mission, one would be obligated to stop fighting at reaching that number of casualties even if the just cause has not been achieved. By contrast, the "prospective proportionality" view, which I share with McMahan (2015),[11] allows that if we expect a war to be beneficial going forward – say doing only 500 more casualties and achieving all the goods of the just cause forecast – then it would now be proportionate to continue fighting. Fabre objects that "[O]n this [prospective] view, proportionality would lose much of its bite as a constraint against killing" (2015, 635–9). If a party were to continue to mistakenly forecast that its war would be proportionate going forward, the prospective view potentially justifies infinite casualties. Moellendorf concurs that discounting past casualties as "sunk costs" leaves "no principled stopping point," whereas "[T]he concept of proportionality assumes ... that there is some limit on moral costs beyond which the continued pursuit of a just cause is not justified. A conception of proportionality which would allow infinite human killing is inconsistent with this" (2015, 665).

On one hand, I would argue that the "infinite casualties" problem should be non-existent in practice, for if a war has been continually more costly than expected, one should adjust one's estimate of costs going forward. Wars that have proven costly with little progress, like Vietnam in the 1960s and 70s and Afghanistan over the past two decades, should be expected to remain so unless there is good reason to expect a change in the near future. On the other hand, the quota view of proportionality has much more counterintuitive implications. For example, if we reached a war's quota of, say, 100,000 casualties without achieving the just cause, and that cause and all the surrounding good that was thought to justify the proportionality quota could be achieved at the cost of one more life, it would be obligatory to stop the war immediately and forego the victory and all its benefits,[12] because continuing would be "disproportionate." The quota view also mistakenly implies that a war will be proportionate so long as it stops short of the initially established quota. However, a war that loses 99,999 lives without achieving any good should itself be called disproportionate. Conversely, if one had achieved 99% of the just cause without any casualties, it would be a mistake to say that one still has one's full quota of 100,000 casualties to spend to achieve the last 1% of benefits. In all of these cases, the prospective approach is more plausible than the fixed quota view. Moellendorf and Fabre want proportionality to establish a fixed, absolute threshold for killing, which its utilitarian, conditional nature does not admit.

Ultimately, I think we have to bite the bullet and say that if there is every reason to think a war going forward would be proportionate (and meet all the other just war criteria), it is proportionate to continue, even though the war as a whole will have been disproportionate. Although there is a sense in which this view discounts the past, it also states that if a war's total costs exceed the total expected value, then it should not have been started in the first place. One can consistently say both that a war should not have been started and that it is justified to continue it now. Indeed the idea of a possible change in a war's moral status was a central insight of the rejection of the one-off view that Moellendorf and Fabre share.

Additionally, as David Rodin (2015) notes, during the course of war, new threats might emerge such that the just cause is more important and there is more to gain by fighting, such that it would be warranted to add to the initial judgment of proportionate cost in order to achieve these expanded benefits. Such shifts in proportionality are an additional reason to adopt a prospective rather than a quota view.

I am sympathetic to the caution raised by skeptics of the prospective view of proportionality. Rodin warns that there can be "a war trap," in which one begins a war based on an underestimation of costs, but then it is hard to stop that which we shouldn't have started. I concur with his conclusion that "the barriers to commencing war must be commensurately strengthened" (2015, 695). The best way to avoid being trapped in a war is to not go in the first place. However, and to the point of this chapter, this is also reason to maintain a high threshold for the continuation of conflict, ending it if there is good reason to think it will not meet any of the criteria, including proportionality, going forward. If one adopts skepticism about war continuation, then the "war trap" loses some of its grip.

There is a powerful psychological and political reason that wars tend to be continued even after they are apparently unwinnable or otherwise foreseeably disproportionate. When a party has already invested substantial resources, including lost lives, stopping the fight makes the costs appear wasted. By contrast, as long as the war is continued and one holds out for an eventual victory, prior sacrifices were not necessarily in vain. Closing the book finalizes the loss and disproportionality. For political decision-makers, ending a war admits defeat and is thus likely to be seen as weak and be nationalistically condemned even if indicated by moral reason. Notably, McMahan argues that redeeming the lives of those lost actually can be a good reason to continue war: "if the just cause for which [a combatant] sacrificed his life is achieved ... his loss is partially redeemed" (2015, 711). He concludes that this redemptive value can contribute to the proportionality of continued war and even become part of the just cause of the fight going forward.

The idea that a duty to the fallen can give an obligation to continue fighting an otherwise unjustified war until victory is interesting, but I think largely misguided. I reject the view that says we lack duties to the dead because they have no interests; that involves an implausibly rigid moral ontology. Intuitively, respect for the former individual and a sense of solidarity give us obligations to the dead, such as to respect their wishes for the disposal of their bodies and property. This duty can help make a costly endeavor proportionate, such as scattering a person's ashes in a remote location per their request. In this vein, it is clear that a community owes something to its war dead, at least a commemoration of their sacrifice on its behalf.[13]

However, it is not clear that it honors the dead to continue to fight for their cause when that mission would not otherwise be morally justified. One might say that it would more properly honor them if we learned from their loss and refused to repeat the mistake. This is particularly true if the cause for which they died or for which the party now fights is unjust. Indeed McMahan hastens to add that a war has no redemptive value if it lacks a just cause. However, I would argue that it is also not redemptive if it is no longer a last resort, properly authorized, or proportionate in cost.

Continuing to fight for an unjust or unachievable (or not justly achievable) cause isn't something that the dead would typically or should rationally expect of their comrades. If the cause is winnable but disproportionate in the amount of destruction (in terms of civilian casualties as well as the loss of combatant life), then one might achieve the mission one shared with one's comrades but at an excessive moral cost. There are ways – public commemorations and personal remembrances – to honor those who fought and died without fighting a costly war until either victory or decisive defeat. Even if not giving up their fight was an important way of honoring the dead, it would be insufficient to outweigh the duty to promote the welfare of the living. If we promise to carry out the last wishes of a relative but find that these are immoral, the other moral duty either nullifies or outweighs the duty of fidelity to the deceased. If this is correct, we should be skeptical about the idea that redemption of the dead could provide a just cause or proportionality to additional fighting and killing.

McMahan's argument is based in part on his view of the moral *inequality* of combatants, wherein only those fighting on the just side have a right to kill. From this perspective, if one's side is just, every casualty suffered is a new injustice committed by the opponent. This, McMahan suggests, gives us additional reason to fight the ongoing perpetrators. This view is problematic for several reasons. First, as detractors of the moral equality view note, there are many arguments that at least excuse, if not justify, fighting on the wrong side.[14] "Unjust combatants" should not be viewed as murderers, and even their deaths are an evil to be minimized. A second problem is that since, as is widely acknowledged by students of war, each side usually thinks its cause is just, any additional rights to wage war granted to the just side will be taken by both. Thus, treating redemption of the deceased as grounds for continued fighting will encourage the continuation of wars without just cause as well as unnecessary or disproportionate fighting for just causes. McMahan qualifies that a war should not be fought *solely* for the cause of redemption, as this would amount to a war of vengeance (714). However, insofar as he allows avenging deaths to tip the balance toward just cause and proportionality, McMahan still gives vengeance a role in justifying wars. There may be a modicum of value in meeting out punishment and signaling respect for those who have died fighting for our (we think just) cause. Yet, as McMahan himself has argued, war is not an effective tool for administering punitive justice, as it kills the innocent and relatively innocent along with the guilty (2008).[15] This limited retribution is not sufficiently valuable to justify the killing of more innocents. If the redemption of those lost justified continued fighting, it would imply a potentially endless cycle in which those killed in the course of vengeance also have to be avenged. For all these reasons, the limited value of expected redemption cannot justify the continuation of a war that is not otherwise justified.

7.4 *Ex bello* concerns: the case for continuing the fight beyond *jus ad bellum*

As I noted above, Moellendorf and others theorize the ethics of ending, and by implication, continuing wars under the rubric of "*jus ex bello*."

Fabre argues that whether to end a war is an ethically different decision from whether to start it, concluding that "we must sever the ethics of war termination from the ethics of war initiation" (2015, 652), a view that Moellendorf in concurrence dubs the "independence thesis" (2015, 654). While these theorists agree with my view that a war's justification can change, they do not consistently condition the continuation of war on its satisfaction of the full range of *jus ad bellum* criteria. The term "*jus ex bello*" and Rodin's (2015) alternative "*jus terminatio bellum*" both imply that what is to be justified is concluding a war now underway. By contrast, I emphasize that the presumption against war implies that its continuation should be problematized. If one finds oneself fighting without just cause, proper intent, legitimate authorization, proportional results, last resort, or likely success, one ought to stop one's fight, just as one ought not to have started fighting under such conditions. The *jus ad bellum* criteria are those that are needed to justify fighting a war at all and thus should be applied at any point the continuation of the war is in question.[16]

Moellendorf initially lists the following *ex bello* principles: "just cause, proportionality, likelihood of success, and [instead of last resort] the pursuit of diplomatic remedies" (2008, 134). This reveals two additional problems with his "*ex bello*" terminology. First, the conditions listed are not criteria for exiting war but rather for justly *continuing* it. The second thing to note is that the "*ex bello*" criteria are identical to the *ad bellum* criteria, illustrating that it is not a distinct realm of moral reflection.[17] Why this is true is clear if one does not use the misleading phrase of "*jus ex bello*" and referred to *jus ad continuandum bellum* or, more simply, "*jus ad bellum.*" The issue is one of just recourse: whether it is justified to be at war at all.

Moellendorf goes on to add a rule which is indeed a condition for war exit,

> the principle of moral costs minimization: A war should be ended in a manner that minimizes costs that arise in the process of ending, especially the moral costs to civilians, the institutions of a just and peaceful social life and the country's natural resources and vital infrastructure. (2015, 670)[18]

A troop withdrawal could have foreseeable harmful effects, leaving a region in chaos or under the control of or susceptible to a repressive or murderous force. Moellendorf concludes that "the path to peace in an unjust war [which violates *ad bellum* principles] might itself be unjust" (2008, 135). Later, Moellendorf describes the "costs minimization" principle as in tension with a competing principle of "all due haste," that a war should be ended as soon as it stops meeting the *ad bellum* criteria. Moellendorf suggests that in a case like Iraq or Afghanistan, the costs minimization principle helped to justify, indeed required, staying after *jus ad bellum* was otherwise forfeit, overriding "all due haste."

The case for continuing a war that no longer satisfies jus *ad bellum* has also been made by Orend, albeit under the rubric of "*jus post bellum*" instead of "*ex bello.*" Although *post bellum* literally means "after" war, in Orend's work, it includes matters of finishing the war justly. Finishing justly, in turn, includes ending the war at the appropriate time, which centrally involves not staying and fighting too long or leaving too soon. Even though Orend (2000) emphasizes that justly finishing the war is related to justly starting it, he does not require the application of the full array of *ad bellum* principles as I defend. "The proper aim of a war is the vindication of the rights whose violation grounded the resort to war in the first place" (123). That is, one should fight until one achieves one's justifying cause, the protection of rights through national defense or humanitarian intervention. In his subsequent work, Orend (2013; 2014) adds a responsibility to remain to rehabilitate a conquered state, even if the victor was initially an aggressor.

Although Orend's *just post bellum* retains a just cause criterion related to *jus ad bellum*, his discussion defends elements of the one-off *jus ad bellum* view. First, he doesn't include requirements for last resort, proportionality, authorization, and likely success in the continuing prosecution of a war and its occupation stage. This implies that a well-intended war, once begun, can be justly continued even if it is clear that it will be harmful on balance, unwinnable, illegal or otherwise inappropriately authorized, and if there are feasible peaceful alternatives to war. Orend also would set a lower just cause bar for continuing an intervention than for starting it. For, in his discussion of just cause for humanitarian intervention, Orend concurs with Walzer that "the human rights violations in question must be 'massive' and 'terrible,' such as incidents of 'massacre or enslavement' in order to justify armed humanitarian intervention." However, when subsequently defending the forcible reconstruction of a conquered state, Orend argues that "the goal of justified post-war regime change is the timely construction of a minimally just political community" (2013, 221). Continuing intervention becomes justified without the massive and terrible rights violations required to intervene initially. On these grounds, Orend defends his view that the US and its allies are not only justified but also obligated to continue to forcibly occupy Iraq and Afghanistan even if this is not necessary to stop an atrocity.

While Orend does not explicitly defend this discrepancy between the criteria for entering and continuing wars, a few possible arguments are implied. He argues for a presumption of regime change and rehabilitative occupation on the grounds that human rights are not secure without a stable institutional resolution. A war that simply stops gross human rights violations without removing the underlying conditions that gave rise to them may see the violations repeated. "Reconstruction needs to take place within a secure context, and the war winner is clearly best positioned to provide this" (2013, 232). This is a utilitarian consideration in line with Moellendorf's "cost minimization." There is also a deontological argument for a continued military force beyond *jus ad bellum*. By having become involved in the fight, one bears responsibility for the situation that one leaves. This responsibility to provide continuing security is, if anything, heightened for an aggressor who invaded wrongly. Orend endorses the "Pottery Barn Rule": "having broke it the war-winner bought it, i.e. shouldered the responsibility for the reconstruction of a replacement regime" (232). This expresses an argument for remaining regardless of overall utility calculations as well as the satisfaction of the other just recourse criteria. I will respond to the second, deontological, argument before taking up the consequentialist one.

7.5 Incurred responsibility and the ethics of war continuation

It is correct to describe war termination as a new decision and one with considerable consequences and thus responsibility. However, it does not follow from this that war continuation is presumptively justified. War, with its killing and mass coercion, has the burden of justification. The same considerations imply that it will be wrong to continue a war if it does not meet the criteria for just recourse. By contrast, stopping fighting does not directly violate rights and does not require a particularly weighty justification. It only requires that there is not an overwhelming moral case to fight.

There is a difference between first deciding to enter a war and later deciding whether to continue it. One's previous determination, presumably seriously undertaken, of a war's justification, gives reason to presuppose its continuing justice so long as no new considerations call the justification into question. However, if there are significant changes in circumstances, information, or mission, as discussed above, then this presumption of continuation no longer holds. One might also realize that one's previous decision was faulty, so the presumption of continuity, while of some occasional practical force, cannot bear justificatory weight.

There are two ways that responsibilities might be heightened in the course of war. One is that the war party is responsible for having brought about the suffering of those in the area which it occupies. This gives it a particular obligation to alleviate these conditions. This obligation to mitigate war harms is, seemingly paradoxically, even greater if one's initial war was unjust. For example, the US was arguably obligated to stay to help reconstruct Iraq not just despite, but also because of, having invaded unjustly. It seems odd that a wrongful invasion can create a right to stay to repair it. I suggest that a right to continue fighting or occupy a territory can only develop if the new war intends the just cause, is necessary, proportionate, and properly authorized. These should be applied with particular skepticism in the case of a recent aggressor.

Even a just intervener may acquire obligations to rehabilitate. Any use of force will have caused harm, including harm to innocent bystanders. It is doubtful that one is responsible for reparations for casualties to enemy combatants who were liable to attack, though belligerents are responsible for administering to the enemy injured as well as their own. Parties bear particular moral responsibility for redressing harms to non-combatants as well as any unnecessary or disproportionate harm to combatants. While the actions which harmed non-combatants may have been justified through double effect reasoning, the collateral damage to undeserving victims is not itself justified. Just as there is a duty to minimize such damage, avoiding it if possible, there is presumably a duty to redress this unjustified harm. Such a duty falls at least partly on those whose actions caused, and may have benefited from, the harm. This duty can include aid in the reconstruction of civilian infrastructure that has been damaged in the course of one's war, as well as providing relief to the injured, displaced, and otherwise aggrieved, such as victims' families. The invader's just cause of upholding its own rights or protecting others in a humanitarian intervention does not waive its obligations of reparation. Indeed it would be particularly contradictory for a supposedly humanitarian intervener to deny an obligation to alleviate suffering and rights infringements for which it bears the responsibility (Pattison 2014, 129; Coleman 2013, 201).

There are questions about to what extent these responsibilities fall specifically on the intervener. Arguably, the vanquished enemy retains responsibility for alleviating the conditions of its war injured, families of the deceased, and damaged infrastructure. The vanquished home state would be especially responsible if it were the aggressor in the conflict and gave just cause for the enemy's attacks upon it. Even if the home state was wrongly invaded, some of the harm to be reconstructed after an invasion may be caused by other parties, including insurgents and remnants of the government, as was the case following the invasion of Iraq.

In his rejection of the "Pottery Barn" "you broke it, you fix it" rule Michael Blake (2014) argues that the broader world community bears some responsibility for reconstruction, regardless of intervener culpability. Blake is surely right that other parties do not lose all responsibility to assist. If an aggressor is not willing or able to stabilize an invaded territory, or could do it more quickly and effectively with help, then other parties acquire obligations. Yet Blake adds that an intervener only has particular responsibility for repairing the harms it directly caused, and not blanket responsibility for providing security and reconstructing infrastructure damaged by other parties. In Iraq, he concludes, the duty to alleviate suffering not caused by the US falls on the Iraqis themselves or is dispersed through the world community (135–9).

Although I am skeptical about open-ended rights to reconstruction, I think Blake underestimates aggressor responsibility. To the extent that harms caused by insurgents were a foreseeable side effect of its actions, the intervener has responsibility for mitigating these as well as those it directly causes. Culpability and reparation duties can be overdetermined, such that several parties share a duty to assist. Beyond culpability, it may also make sense to hold war wagers strictly liable for reconstruction, which Blake acknowledges could beneficially dissuade parties from reckless intervention (140–7). I would add that strict liability spreads the costs from the victims to those who are relatively capable of paying. It also spares the need to prove each case of fault liability in a suit against a powerful party. It is frequently accepted that if one is engaged in an inherently dangerous activity, one can be held responsible for any damages even if one was not negligent in causing them. For example, if a car driver goes into an unavoidable skid and harms a pedestrian, the driver acquires particular responsibility for attending to the victim and compensating for harms, despite lacking moral culpability.

For my thesis, the question is whether the responsibility to repair or mitigate injustices can translate into a right to continue to wage a war that otherwise fails to satisfy *jus ad bellum*. I argue that the responsibility does not waive or override the prohibition against fighting without just cause, last resort, and proportionality.

To the extent that we have intuitions that one is obligated to fight a war that seemingly lacks *jus ad bellum* in order to prevent injustices, it is probably because one judges that these further injustices actually tip the balance in favor of *jus ad bellum*. This is most straightforward with proportionality: harms that might result when one leaves the war might make continuing an otherwise disproportionate war proportionate.[19] If threats make a diplomatic resolution overly risky, then this, in turn, can make continued occupation and war a last resort, *prima facie* alternatives notwithstanding. Finally, the prevention of injustice, if severe enough, could itself amount to a just cause for war when other missions would not justify continuing it. Orend and Moellendorf's *ex bello* concerns are among the factors that could change a war's status in terms of satisfying those criteria. For example, the US arguably gained a just cause for occupying Iraq after its initial aggression in order to prevent a humanitarian catastrophe for which it would have borne responsibility (Rocheleau 2010a). If continuing intervention was justified in Afghanistan following the dismantling of the Al Qaeda forces there, it is probably by a duty to prevent human rights violations in the unstable new climate. According to my thesis, however, continuing the war is only justified if it would have been initially just to go to war for the same reasons.

If the circumstances do not satisfy the just recourse criteria, it does not make sense to say that war should be continued. If there is not an injustice of the sort that could justify war, or if there is an available feasible diplomatic solution short of war, or if fighting would be disproportionate or unauthorized, then one should stop fighting, even if this would in certain ways involve a failure of complete fulfillment of *prima facie* duties of reparation.

One might wonder whether the responsibility acquired mid-war implies that there is a lower just cause threshold for *jus ad continuandum bellum* than *jus ad bellum*. Arguably one might say that the conditions in Iraq between 2003 and 2010 would not have warranted a new military intervention but did warrant the US remaining due to its responsibility for the new crisis. I would say two things about this. First, I am not sure that there is a lower threshold, such that the Iraq case both did warrant continuing intervention and would not have warranted outside intervention. If responsibility for the problem gives a belligerent in the midst of war additional rights to intervene, amounting to a just cause, then similar responsibility would presumably provide cause in a *jus ad bellum* decision at the beginning of a war. If, without having already deployed forces, A's actions harmed neighbor B, causing B's government to collapse, and this lead to a threat of chaos in B, A could acquire cause for humanitarian intervention by virtue of its responsibility. If this is the case, then while responsibility is efficacious, there is no discrepancy between *jus ad bellum* and *jus ad continuandum bellum*. Secondly, I take it that in the example, the case for intervening at a lower threshold due to one's past aggression is dubious. While A has responsibility for the problem that has arisen, this very responsibility also calls into question its legitimacy as an intervening force. I am inclined to say these conflicting responsibilities cancel out and do not suffice to waive or reduce *ad bellum* war conditions. If forces were justified in continuing wars in Iraq and Afghanistan, it is because the situations met the threshold of atrocity prevention.

There is a second manner in which just interveners and, somewhat more problematically, unjust interveners can acquire a duty to continue their struggle beyond their initial *ad bellum* justification. By virtue of acquiring a monopoly on power, they are obliged to maintain justice in the territory they control. Legal requirements are stipulated in the Hague Convention of 1907:

> The authority of the legitimate power having in fact passed into the hands of the occupant, the latter shall take all the measures in his power to restore, and ensure, as far as possible, public order and safety, while respecting, unless absolutely prevented, the laws in force in the country (Article 43).

The duty to protect the rights of those under one's governance might override the *prima facie* duty not to engage in violence and interfere with local self-determination. An intervener's abrupt departure could leave an anarchic power vacuum or allow domination by human rights abusers. While the threat of these harms might not have been large enough to justify an initial invasion, the war winner's newly assumed responsibility might obligate it to stay, *pace* Orend.

I would again reply that this does not contradict the continuous applicability of *jus ad bellum*. First, much of what is intuitively accepted as justified in an ongoing occupation is policing actions that do not require the same justification as a war. If the occupation's authority is generally accepted and there is no warring party struggling for control of the government, then its stability-operations are no longer acts of war and thus no longer require an *ad bellum* justification. Of course, one should be suspicious of colonial claims to acquire legitimate authority. Acceptance may be forced upon reluctant locals who have no other choice or achieved through propagandistic indoctrination, both of which reduce a right of enforcement. Local support of an occupier is apt to be mixed at best with continuing resistance. If the occupier is substantially unwanted or is fighting a conflict against organized parties that go beyond mere criminals, as was the case with the US occupations of Afghanistan after 2001 and Iraq from until at least 2009, then it is appropriate to speak of continued *war*, and the *jus ad bellum* criteria apply.

In Chapter Four, I argued for a presumption of the legitimacy of those in power due largely to stability concerns. One may wonder if my presumption against occupation contradicts my argument there.[20] In response, a power can lack legitimacy if there is large-scale opposition or another body rivals it for recognition. Moreover, legitimacy justifies police measures but not necessarily war measures against rival groups.[21] If this is correct, a right to use force during occupation takes the form either of (a) preventing threats of massive human rights violations as in standard *jus ad bellum* or (b) an exercise of police force short of war by virtue of being the recognized and capable authority on hand. Neither involves a right of war continuation at a lower threshold than for initiation.

7.6 Utility considerations for continuing a war

In addition to considerations of accrued responsibility, there is a utilitarian case for continuing war at a lower threshold than the initial *jus ad bellum* criteria. Stabilization to prevent or undo harm post-war has practical value as well as principled duty behind it. The fact that one's army is already deployed and engaged means that continuing to fight now is relatively efficient compared to redeploying in the future. The concern for the good that one could do and the value of doing it now as opposed to returning later are clearly relevant in arguments for a presumption of war continuation.

It is generally recognized, and I have maintained in this text, that utility alone cannot justify waging war. The proportionality of benefit to harm is essentially a utilitarian requirement, but it is just one necessary *jus ad bellum* condition and not itself sufficient. One must have a just cause, be properly authorized, and be fighting as a last resort in addition to expected benefits. However, my pluralism allows for a role of utilitarian – especially rule utilitarian – considerations in conceiving of just cause and other *ad bellum* criteria. Thus, I am committed to taking into account the general consequences of a principle of presumed war continuation as opposed to war exit.

Daniel Statman emphasizes the value of achieving a decisive victory against an aggressor. If one stops as soon as one has repelled the enemy but does not continue to fight to destroy their forces, this could allow a new attack later. To crush the disposition to aggression along with its immediate material conditions, there can be value in carrying on a fight beyond the repulsion of an attack. Statman advocates a social contract view of the end of war, in which parties would agree to end war short of entirely destroying the enemy but which "would exempt the parties from the requirement to make sure anew every day that the war is still necessary to achieve the political aims that justified launching it in the first place" (2015, 738).[22] Thus, I take his view to be permissive of war continuation on utilitarian grounds though it may not satisfy just cause and last resort.

Even Walzer defends a certain right to carry on a war beyond repulsion of aggression to the point of "reasonable prevention" of its repetition. If the defeated foe is "Nazi-like in character" then the aggressor is subject to wholesale reconstruction. For most vanquished rivals, Walzer defends limited dismantling of military capacity and reconstruction to ensure "a better state of peace ... more secure than the status quo *ante bellum*" (1977, 121). These rights are based at least in part on long-term consequences.[23]

There are also efficiency concerns regarding ending war that have to do with the costs of deployment. Transporting troops to the battlefield is costly. To withdraw and then have to redeploy them would multiply these expenses and result in a strategically and morally costly loss of time in engaging in the new conflict. Additionally, there are political costs of war entry in terms of convincing home and international publics and securing the necessary votes and support to deploy forces, which may have to be spent anew if one withdraws. While a continuing war can face criticism (and I have argued justly so), in practice, it is unlikely to face the same challenge of justification and criticism that a new deployment would.

Historical cases lend support to the idea that it is better to carry on war until an enemy or injustice is thoroughly defeated rather than to stop fighting early as soon as a tolerable point is reached. After the first Gulf War ended with a refusal to pursue the Iraqi aggressors back to Baghdad and remove Saddam Hussein from power, the United States returned 12 years later with a new coalition of allies. The delay meant that Iraqis had to live for an extra dozen years under Hussein's rule. It also multiplied the expenses of allied forces, who probably could have overthrown Iraq more readily when the latter's forces were in disorganized retreat from Kuwait in 1991. However, it is impossible to predict with precision the costs of a hypothetical invasion, especially the relative ease of the reconstruction task, which was the stumbling block for the later, actual intervention. This example is further weakened by the dubiousness of the 2003 invasion, which many would judge to not itself have had a just cause or been the last resort. The fact that the US unjustly invaded in 2003 cannot retroactively justify invading in 1991.[24]

World War I is another famous case where a war terminated before engaging in the thoroughgoing reconstruction of the defeated state. This permitted the rise of Nazism and a much more costly Second World War. In retrospect, it appears that the Allies should have sought a more decisive overthrow of Germany and its militarism at the end of World War I. A similar case might be made for having continued the Korean War until the

North was overthrown, as the latter dictatorship and its tense border with South Korea have been an ongoing threat to human security. There is historical evidence of the dangers of leaving a bad actor in place. More recently, the US withdrawal from Iraq in 2010 permitted the emergence of the brutal Islamic State, and its withdrawal from Syria in 2019 permitted Turkish incursion against the Kurds and the solidification of power by the Assad regime and its Russian supporters. Withdrawing forces can lead to harm.

However, as with the argument for preventive wars, this utilitarian argument is flawed. One does not know in advance that there will be a continuation of aggression or a restart of humanitarian abuses. To fight now in order to avert a future threat replaces a possible war with a certain one. Moreover, the effects of war are hard to predict. It isn't necessarily the case that fighting longer will get one closer to one's goals, as it may lead to greater resistance and lost allies and legitimacy. As we saw in Chapter Five, state-building has proven difficult, even for the sole superpower, and intervention can play a destabilizing role as well as the stabilizing one that Orend expects. It is thus far from straightforward that prosecuting a war until the complete overthrow of the aggressor and the establishment of a minimally decent state is going to be beneficial. Although its benefits are uncertain, the logic of fighting until regime change and comprehensive rehabilitation, *pace* Orend, will certainly prolong wars, most of which end without regime change and occupation. This policy would have further prolonged wars such as Vietnam, Revolutionary, and Mexican-American Wars, and the War of 1812, as well as World War I and the Korean War, to use US examples. In recent years it would have meant a prolonged mission in Somalia to defeat the warlords, occupation of Libya after the 2012 assistance to the rebellion to overthrow Gaddafi, perhaps reconstructions of Pakistan and other states where drone strikes have been common, as well as prolonged intervention in Afghanistan and Iraq.

These examples show that a decisive, stable peace is a dangerous ideal to pursue militarily. Ending wars when there is not a decisive moral cause or fighting is no longer a last resort is akin in importance to avoiding beginning such a war. I would second the argument of Gabriela Blum and David Luban that it is excessive to insist on fighting until there is no risk whatsoever of continued enemy aggression. Using the war on terror as an example, they point out that an insistence on complete victory implies a never-ending war. One has to settle for a tolerable degree of risk or, as Blum and Luban call it, a "morally justified bearable risk" (2015, 768). They note that we live with many sorts of risk – including by not using force preventively against threats in the first place – and challenge as inconsistent our severe intolerance of risk in considering whether to end a war. This further specifies a limit to the content of Walzer's "reasonable prevention" in *jus ad continuandum bellum*. The value of ceasing a war that no longer has a just cause or is otherwise unjustified, thus avoiding endless war, outweighs the cost of a degree of added risk with a possible need to redeploy in the future. If valid, this analysis of tolerable risks is consistent with my view the presumption against going to war applies to its continuation as well.

My position that wars should be ended as soon as they cease meeting any of the *jus ad bellum* criteria may seem hopelessly impractical and even dangerous. Do we not need to take into account Moellendorf's "cost minimization"? In response, several considerations show how "all due haste" can be effective and minimally risky. First, potential aggressors can be contained without continuing the war against them. Forces can be retained nearby in non-combat modes, ready to redeploy relatively quickly and inexpensively if the need – and justification – arises. Containment appears to have been largely effective in the cases of North Korea and Iraq, as well as other potential aggressors like the former Soviet Union. Second, it is important to remember that assistance to those in need does not have to take a military form so that it is a false dichotomy to say that one has to either continue a war or do nothing in the face of injustices and bad consequences. Third, while I have said that one should stop a war when one no longer has a just cause of countering aggression or human rights abuses and when there is a resort other than war, this should not be interpreted to mean that war must be ceased against parties whose intentions, institutions, and power imply that they will continue or immediately resume their injustices. Disarmament of an aggressor and overthrow of a regime whose existence is an imminent threat to human rights is an extension of the causes of resisting aggression and undertaking a justified humanitarian intervention. I would agree with Walzer that the occupation and reconstruction of Germany (and probably Japan) after World War II was justified to put a stop to their aggression and atrocities.[25] However, it is questionable whether the long-term sovereignty-infringing occupations were justifiable for their duration.

Probably the most likely case for just war continuation in violation of *jus ad bellum* principles is a situation in which continued belligerence has a just cause and would be proportionate, but there is a possible path to peace such that fighting is no longer the last resort. It is such a situation in which "cost minimization" could be most plausibly said to trump "due haste" in war termination. Arguably this was the case in Afghanistan and Iraq when there was debate about whether to negotiate with insurgents. In my view, elaborated in Section 7.8, the continuing validity of last resort means that such measures must be pursued if feasible. This is the case even if one thinks that war might be a faster or more certain means of achieving a just result. First, killing should be avoided if possible, and, secondly, from a rule utilitarian perspective, war tends to be more costly and less effective than predicted.

7.7 Liability and the justification of war over time

In countering utilitarian rationales, I just argued that if there is not an ongoing large-scale rights violation, then war is not justified. However,

Thomas Hurka (2007) argues that if there had been an initial rights violation by a state or group of actors, this can create a liability to attack, which continues even after their rights violation is stopped. As we saw in Chapter Three on punitive war, Hurka holds that a basic rights violator is "globally liable" to attack for additional "conditional" purposes beyond defending against its assaults. For example, Hurka argues that in World War II, the Allies justifiably fought to disarm and rehabilitate Germany and Japan even after the Axis powers' aggressive conquest had already been successfully overturned. Thus Hurka provides an argument justifying war continuation without the satisfaction of the *ad bellum* requirements.

To counter this argument, I echo my critique of punitive war. A past injustice is not sufficient to justify military force. Although retribution can be a just goal to pursue by means short of war, such as ordinary policing, punishment of wrongdoers is not sufficiently important to justify the large-scale killing, coercion, and destruction involved in war. Without a current justifying cause, the party would attack unjustly and not as a last resort even if it had a proportionate cost-benefits ratio. There can be no global liability making a party liable to any and all coercion that is thought to be for the greater good. This is particularly true when it comes to warfare which harms not only the guilty but also innocent bystanders. The addition of prospective benefits cannot make it a just cause to kill in response to past injustice. Nor, on the flip side, can the past injustice committed make fighting for utility a just cause. Both are category mistakes and, if permitted in just war thinking, would invite a dangerously easy justification of war by reference to a combination of past grievances and prospective future goods.

What of the intuition that it was justified to fight Germany and Japan until their unconditional surrender and then occupy and reconstruct their states? In my view, this can be accounted for without a presumptive right to continue wars that previously had but now lack just cause and last resort necessity. First, the overthrow of the governments was arguably necessary to complete the Allied defense against Axis aggression and the ongoing human rights violations perpetrated by those regimes. The two societies' recent and continuing actions and degree of militarization and military capacity suggested a likely continuation of aggression and internal atrocity crimes if they were left in power. This justified at least an initial period of occupation after the war. Once there is no longer a serious threat of rights violations, occupation by military force is only justified if legitimated by invitation by a representative local government. To the extent that the ongoing occupations of Japan and Germany were forcefully imposed without just cause, they became unjustifiable. The paternalistic element of the Marshall Plan was not, in my view, sufficient to justify indefinite intervention. As I noted above, following Walzer, aggressors and rights violators can be incapacitated and deterred through defeat, such that systematic overthrow and reconstruction of their states are not generally necessary.

In the case of Afghanistan, the initial 2002 intervention was justified as defense against Al Qaeda terrorism following the 9/11 attacks. However, as Al Qaeda was dispersed and distance from the attacks grew, the US justification shifted from defense to humanitarian intervention. The continuing mission was largely to prevent the Taliban and other human rights-abusing groups from dominating the still unstable state. A new mission cannot be rejected out of hand. However, for it to justify continuing war, the cause must involve the prevention of atrocity level rights violations of the sort thought to warrant armed intervention. The past cause of defending against Al Qaeda aggression cannot make Afghanistan liable to the ongoing war for all purposes deemed good, such as fostering democracy and women's rights. Thus, just cause for continuing the war has been dubious.[26] At the same time, alternatives for negotiating peace with Taliban forces were long eschewed so that the continuing US combat operations against them also arguably violated last resort. As I write this, the US has just undertaken negotiations regarding its withdrawal with the Taliban, conditional on Taliban forces negotiating a peace treaty with Afghan government forces. Since the possibility of peace is sufficient reason to require that war end and is necessary to end most wars, I turn to the ethics of treaties.

7.8 Ethics and negotiating peace

Although parties can unilaterally withdraw from conflicts, typically, wars end with a peace agreement. This includes a mutual commitment to stop fighting as well as terms on matters such as war crimes trials and reparations. Treaties may also place conditions on one or both sides regarding territorial rights, armaments, and national laws going forward. There are moral questions surrounding when peace negotiations are legitimate and obligatory and what their ethical limits are.

On the one hand, the last resort criterion implies that one should accept a minimally decent peace offer rather than continue to wage war. The question arises when a deal is acceptable such that one is obligated to take it. Clearly, if a peace treaty could fully achieve the just cause, it would be obligatory to accept it. However, it is also required to accept a peace deal that partially achieves the just cause or simply minimizes losses without achieving it at all, if this is most just on balance, with acceptable risks. Even if one thinks one could achieve the cause more fully by fighting, one should stop either if doing so would be disproportionate or if fighting were thought proportionate, but peace could be had now with tolerable risks.[27]

As Walzer argues, a too-ready peace "may confirm the loss of values, the avoidance of which was worth a war" (1977, 123). It would not be required, and indeed it might be morally unjustified, to negotiate peace without a cause attained or with a cause only partially attained if continued fighting were necessary to stop large-scale injustices of atrocity or aggression and could do so at proportionate cost.

Lazar (2017) proposes two principles of just treaties and war conclusion, which Moellendorf embraces as well. First, parties should exercise "good faith," making sincere commitments and adhering to them (236–7). Launching attacks after having promised to stop fighting would be a form of perfidy, unethical in itself, and undercutting trust and thus the possibility of negotiating peace. "Perfidy in *jus ex bello* undermines not only a particular peace, but the prospect of peace after war in general" (237). The sincere negotiation principle has the corollary that a party is not obligated to accept a treaty if it has good reason to believe that the other side is not negotiating in good faith. This raises the problem that one can never be sure in advance of the good faith of one's enemy. However, a combination of indications of sincerity and mutual interests in securing an eventual peace can make it rational to accept negotiations in good faith.

Problematically, notoriously untrustworthy parties may not be easily trusted to negotiate in good faith. Statman argues that due to the untrustworthiness of their promise, wars with terrorist groups will have to be carried on until those groups are decimated (2015, 745–8). While there is reason for skepticism about groups that have repeatedly violated moral norms, it would be a mistake to categorically reject negotiations with types of groups. I take it that historically there have been successful negotiations with groups that practiced terrorism, such as the IRA and PLO, and that practical incentives, as well as perhaps moral concerns for integrity and credibility, made such trust not absurd. Trust in negotiations need not be blind. There can be verification and a threat of the re-taking up of arms or implementing sanctions short of war for treaty violations.

In terms of the content of negotiated terms, Lazar recommends a second principle of "no new rights": parties are only entitled to fight "to restore rights that were justly held before the war, not to create new rights" (2015, 237). Even if a party has the upper hand in negotiations, it should not try to take advantage of this to impose terms that are clearly unjust. For example, it would be wrong to negotiate for harsh collective punishment, which falls upon innocent as well as guilty. It would also be wrong to attempt to gain rights to control land and people beyond temporary measures necessary for the purpose of stopping aggression or atrocities.

There are complexities in the application of the "no new rights" principle, which require its contingency. To say that no side should propose unjust measures does not mean that it might not be obligatory for the other side to accept them if no more just or proportionate solution is possible. A negotiated surrender may be the most just of the unjust conclusions within reach. There is tension between the principles of good faith negotiation and no new rights in treaties. This is most extreme in the examples just given, where a party has agreed to terms that are clearly unjust. Here the question becomes whether the abused party is obligated to observe the terms it conceded to. There is a *prima facie* obligation to keep one's promises, avoiding perfidy in particular, and adherence to the norm of keeping treaties is required for their continued effective use to secure peace. However, there is also a *prima facie* right to resist injustice, and in this case, the terms were unjust and agreed to only under coercion. In some cases, the right to fight injustice will permit parties to break their treaties. For example, we tend to praise the French resistance for continuing to fight despite the treaty that their state signed with the Germans.

On the other hand, every treaty will involve at least one side, if not both, accepting some terms viewed as unjust. Ideally, of course, the just side will have won and only imposed reasonable restrictions, such that no injustice is involved in adherence. However, in reality, the just side may not always win a war, and from the standpoint of the loser, the victor will generally appear unjust in any case. Moreover, even a just victor can impose unjust terms, such as excessively punitive sanctions and occupation. The side which is losing or suffering more in a war is likely to have had to bargain away rights. This may include things such as the right to maintain and develop military power or to retain particular government officials or policies that the other side declared unacceptable. All of this means that negotiation tends to require accepting a degree of injustice, thus according new rights to the other party. Since parties must be taken at their word if they are to negotiate peace and end wars, there will be an obligation to abide by negotiated terms that have a degree of injustice rather than restarting conflict. When injustices of the treaty are not major, the duty of good faith prevails over "no new rights."

7.9 Conclusion

I have argued that ongoing interventions must satisfy the *jus ad bellum* criteria, or their continuation becomes unjustified. It is a mistake to think of the continuation of an ongoing war or imposed occupation as unproblematic due to the previous commitment to use force or even the previous justifiability of that force. In particular, it is not sufficient justification to continue to use force because one expects greater benefits than harms, due to a sense of responsibility or due to an enemy's "general liability" from past wrongdoing. None of these can justify prosecuting a war that fails to satisfy the other *ad bellum* criteria, including just cause and last resort. Parties are obligated to seek and observe acceptable peace treaties.

I have discussed several examples of the problematic justice of war continuation. I turn finally here to the Second Gulf War. In Iraq, there may have been a time at which occupation was justified by the cause of stabilizing the state following the unjustified initial US intervention (Rocheleau 2010a). Orend and Moellendorf defended the continued military mission there with a combination of responsibility and the risks of too early withdrawal. The rise of the Islamic State after the 2010 US withdrawal from Iraq seems to speak in favor of the presumption of continuation and against war exit.

Nonetheless, a plurality of ethical considerations, including principle, pragmatics, and international law, all support observing the norm of not continuing to fight without *jus ad bellum*. Deontologically, armed force needs more justification than a prospect for benefits. The imposition of military force needs a clear just cause and should be a last resort. The mission of combatting remnants of insurgents and policing Iraq without the people's consent could not satisfy the just cause requirement. In terms of consequences, US intervention was not just a solution to but also a catalyst for Iraqi instability. It undermined the legitimacy of the government even as it tried to support it. If it continued against the will of growing numbers, this would be worsened still. The occupation of Iraq against the will of the people would violate and undermine international law. In my view, then, the Obama Administration was required to withdraw troops in 2010 when an atrocity was not imminent, and the Iraqi government no longer approved their presence. However, the US was also probably justified in redeploying troops to fight Isis in 2015. The gap in security and delay in redeployment presumably allowed some additional Islamic State atrocity crimes. Nonetheless, observing international norms and self-determination have value too. Armed force cannot and should not be retained in all areas of the world in which human rights abuses could emerge. As I write this in 2020, the cause for US intervention is no longer strong. Moreover, the Iraqi government is once more asking the US to withdraw such that even policing assistance short of war is no longer justified.[28] I have consistently defended the thesis that continued deployment under changed circumstances must meet standards for *jus ad bellum*, challenging the presumptive permissibility of ongoing force.

Notes

1 Moellendorf (2015) reasserts the framework, and it is accepted by other writers (Fabre 2015) in an issue of *Ethics*.
2 I am grateful to my colleague from Classics, Stephen Kershner, for suggesting this Latin phrase as well as confirming that "*jus ad bellum*" could grammatically refer to rightfully fighting throughout a conflict and not just its initiation.
3 My argument for the continual reapplication *of jus ad bellum* principles over time draws on an earlier article (Rocheleau 2010a), where I argued that continuation of a war such as that in Iraq could be justified even if its initiation was unjust. The concept of reevaluation of just recourse over time in that piece was in turn influenced by Lango (2007).
4 I use "realist" to refer to an approach to war's justification which is restricted to considerations of national interest.
5 A change in mission is essentially a change in intention, so the previous discussion could be viewed as being about the satisfaction of the right intention principle. As discussed in Chapters One and Six, right intention is the subjective side of just cause, so a war without a just mission could also be said to lack just cause.
6 I am not familiar with cases in which the UN has specifically revoked an authorization of war. More common are cases in which a state which initially fought under a UN mandate came to exceed that mandate, for example, in the

2014 intervention in Libya (where force to stop an atrocity in Benghazi was authorized, but the coalition adopted an additional aim of regime change). These are cases of a change in mission rather than a change of circumstances.

7 Moellendorf adds that continued war must satisfy a "principle of the pursuit of diplomatic remedies," so our difference on this point is largely verbal, as he does not recognize this as a continuation of last resort because of his one-off interpretation of that principle (2008, 134).

8 As I discussed in Chapter One, in my view, likely success isn't an absolute criterion but a factor in proportionality. However, since success considerations involve a particular array of factors, it can be helpful to give it explicit independent consideration.

9 Depending on the avoidability and thus culpability of the error, the initial miscalculation may or may not be excusable.

10 One might argue that *all* changes in proportionality and likely success are epistemic. If metaphysical determinism is correct, then consequences are fixed, and it is just one's knowledge of the future which changes over time. However, I follow common sense in not presupposing determinism and assuming that through both the free will of agents choosing new strategies and random changes in conditions and consequences, the future is open. As philosophers since Kant have noted, determinism, with its implications that our choices are inevitable, seems to contradict the idea of moral judgment. Whether or not determinism is true, it remains the case that the future is imperfectly predictable such that expected consequences, and thus judgments about the ethical proportionality of various actions, change over time.

11 I borrow the terms for the contrasting "prospective proportionality" and "quota" views from McMahan (2015).

12 McMahan (2015, 710–7), in his defense of prospective proportionality, argues that the latter does better at valuing the lives already lost by considering their redemption an extra reason to fight. For the reasons I discuss below, I largely disagree with this redemption argument.

13 Whether or not the individual's war deeds helped the community, they lost their lives intending to help it or responding to its orders or request for service. There is an important issue that is outside the scope of this project about whether combatants lost in a (subsequently recognized as) unjust cause should be honored. For example, should Civil War Confederate and World War Two German soldiers be recognized for serving their communities, or should the community refuse to do so to out of recognition of the injustice of its former cause?

14 Many writers have objected to McMahan's case for moral inequality due to its discounting of contextual duties, the pressures upon and epistemic limitations of combatants ordered to war by their states, and the undesirable implications of affirming to each side (which are likely to view themselves as just) that opposing fighters are criminals morally undeserving of traditional P.O.W. rights. As my focus here is upon *ad bellum* issues, I do not take this up in detail in this text. See my review of McMahan, among other critical commentaries (Rocheleau 2010b).

15 McMahan argues that even if the combatants (on the unjust side) who we are currently fighting are not the same ones who killed our comrades in the past, "since he acts wrongly in impeding the achievement of the just cause and therefore also in impeding the redemption of the earlier losses, he can be liable to be harmed as a means of preventing him from being an effective impediment to both those aims" (2015, 715). That it will be redeeming to kill other enemy troops whose liability to attack is contingent on their resistance to one's pursuit of redemption is tenuously circular.

16 One may worry that the requirement of continually meeting the *ad bellum* criteria is too demanding to apply. After all, we cannot expect belligerents to continue to mull over their justification for fighting while they are busy carrying out the fight. My preliminary response is to say that the general moral requirement of continual justification does not have to be consciously applied at each minute. Having recently decided reflectively and not having received any new morally relevant information, we would not need to continually reevaluate. There is also a certain division of labor in thinking through the justification of war. While we do not expect the combatants to continually worry about *jus ad bellum* while they are busy fighting, there are parties removed from conflict who are responsible for and able to make moral judgments. This includes the political leaders, advisors, and citizen scholars, journalists, and laypersons who are more able to reflect on and responsible for answering these questions. Statman (2015, 741–4) argues that the "division of labor" tends to support the presumptive continuation of war, but it is more accurate to view it as limiting who needs to reconsider and how often, but not the need for reconsidering or the criteria to be used. On the other hand, if combatants have a moral obligation to avoid fighting in unjust wars (*pace* McMahan), then they – like political leaders – have an obligation to reconsider their wars if relevant considerations have changed.

17 Moellendorf recognizes that his criteria mirror those of *jus ad bellum*: "Although whether a war should be continued is logically distinct from whether it should be initiated, the considerations that go into answering the former are not necessarily distinct from those of the latter, except in the case of last resort" (2008, 132). The problem is that the over-emphasis on the differentness of "ex bello" logic ends up distorting the need for continual justification and encouraging us to think that war exit needs to meet special criteria.

18 In his earlier article, Moellendorf (2008, 134–6) describes cost minimization as a principle governing "how" troops should be withdrawn. However, since it is a condition for their being withdrawn; it will determine when and whether as well.

19 Seth Lazar (2017) makes a similar point that we need only continue to apply the *ad bellum* principles and do not need different principles for *jus ex bello*.

20 I am indebted to an anonymous reviewer for pointing out this seeming contradiction.

21 While it is beyond the scope of my current project to fully delineate the ethics of occupation and *jus post bellum*, I suggest that these principles of condemning war without an overwhelming cause and rejecting coercion against the will of the people are helpful starting guidelines.

22 I find Statman's contractarian view problematic. It is not clear to what extent it refers to an actual or hypothetical contract. In view of the limits of international law, it seems that it must be in large part the latter. However, the conditions, terms, and force of a hypothetical contract are unclear. Moreover, Statman, following Hobbes, appears to get his normative basis from the mutual self-interest of the contracting parties, which leaves the duty to uphold it uncertain in cases where breaking it appears to be in an individual state's interest.

23 Neither Walzer nor Statman, nor Moellendorf or Orend, for that matter, are utilitarians. However, as elements of their arguments are consequentialist, I take these up in combination here.

24 Below I discuss the fact that the US left Iraq again in 2010 only to return in 2015 to fight Isis.

25 To say that the goals and acts of occupation were justifiable is not, of course, to say that the terror bombing used to attempt to hasten the German and Japanese surrenders was justifiable.

26 In Chapter Six, I critically interrogated the possibility of justifying the Afghan war through a compound cause involving national defense and humanitarian intervention. Compound causes frequently intersect with war continuation ethics, for in the course of a mission, as the initial cause wanes and new possible causes arise, one may wonder about a combined justification for war continuation. As in Chapter Six, my argument is not that the ongoing war in Afghanistan was definitely unjust, but rather that its justification should be tested rigorously according to *jus ad bellum* standards and not given a pass because of its continuity with a previously justified use of force.
27 Related but not identical issues surround the decision about whether to seek and agree to a temporary truce (Eisikovits 2016).
28 With Isis largely defeated, the US is currently more concerned to combat Iranian influence than terrorist groups. While Iranian militias have posed a threat to US troops, self-protection is circular as a reason for staying. If there is an immediate threat to the Iraqi state or US troops were requested for security assistance, this would justify continued US presence.

References

Blake, Michael. 2014. "The Costs of War: Justice, Liability and the Pottery Barn Rule." In *The Ethics of Armed Humanitarian Intervention*, edited by Don Scheid, 133–47. Cambridge: Cambridge University Press.

Blum, Gabriela, and David Luban. 2015. "Unsatisfying Wars: Degrees of Risk and the *Jus ex Bello*." *Ethics* 125 (3): 751–80.

Coleman, Stephen. 2013. *Military Ethics*. Oxford: Oxford University Press.

Eisikovits, Nir. 2016. *A Theory of Truces*. New York: Palgrave Macmillan.

Fabre, Cecile. 2015. "War Exit." *Ethics* 125 (3): 631–52.

Hague Convention of 1907 Respecting the Laws and Customs of War on Land. Accessed October 23, 2020. https://www.loc.gov/law/help/us-treaties/bevans/m-ust000001–0631.pdf.

Hurka, Thomas. 2007. "Liability and Just Cause." *Ethics & International Affairs* 21 (2007): 199–218.

Lango, John. 2007. "Generalizing and Temporalizing Just War Principles: Illustrated by the Principle of Just Cause." In *Rethinking the Just War Tradition*, edited by Michael Brough, John Lango, and Harry van der Linden, 75–95. Albany, NY: State University of New York Press.

Lazar, Seth. 2017. "War's Endings and the Structure of Just War Theory." In *The Ethics of War: Essays*, edited by Saba Bazargan and Samuel Rickless, 227–42. Oxford: Oxford University Press.

McMahan, Jeff. 2009. *Killing in War*. Oxford: Clarendon.

McMahan, Jeff. 2015. "Proportionality and Time." *Ethics* 125 (3): 696–719.

McMahan, Jeff. 2008. "Aggression and Punishment." In *War: Essays in Political Theory*, edited by Larry May, 67–84. Cambridge: Cambridge University Press.

Moellendorf, Darrel. 2008. "*Jus Ex Bello*." *Journal of Political Philosophy* 16 (2): 123–36. 10.1111/j.1467-9760.2008.00310.x

Moellendorf, Darrel. 2015. "Two Doctrines of '*Jus Ex Bello*'." *Ethics* 125 (3): 653–73.

Orend, Brian. 2000. "Jus Post Bellum." *Journal of Social Philosophy* 31 (1): 117–37.

Orend, Brian. 2013. *The Morality of War*, 2nd edition. Buffalo, NY: Broadview Press.

Orend, Brian. 2014. "Post Intervention: Permissions and Prohibitions." In *The Ethics of Armed Humanitarian Intervention*, edited by Don Scheid, 224–42. Cambridge: Cambridge University Press.

Pattison, James. 2014. "Bombing the Beneficiaries: The Distribution of the Costs of the Responsibility to Protect and Humanitarian Intervention." In *The Ethics of Armed Humanitarian Intervention*, edited by Don Scheid, 113–29. Cambridge: Cambridge University Press.

Rocheleau, Jordy. 2010a. "From Aggression to Just Occupation: The Temporal Application of *Jus Ad Bellum* Principles and the Case of Iraq." *Journal of Military Ethics* 9 (2): 123–38. 10.1080/15027570.2010.491325

Rocheleau, Jordy. 2010b. "License to Kill." *Review of Jeff McMahan's Killing in War. Radical Philosophy Review* 13 (2): 203–8.

Rodin, David. 2015. "The War Trap: Dilemmas of *jus terminatio*." *Ethics* 15 (3): 674–95.

Statman, Daniel. 2015. "Ending War Short of Victory: A Contractarian View of *Jus Ex Bello*." *Ethics* 125 (3): 720–50. 10.1086/679561

Walzer, Michael. 1977. *Just and Unjust Wars*. New York, NY: Basic Books.

8 The nature and necessity of legitimate authorization

8.1 Introduction

The focus of this book has been the critique of ways in which the just cause principle has been weakened by recent interventionist arguments. I have argued that punishment, most human rights violations, most compound causes, proportionately small injustices, and formerly justified causes should not be accepted as just cause for war. Despite the breadth of various erosions of the just cause concept, the most contested just war principle may be that of legitimate authority. Although the authorization requirement dates to the early just war theories of Augustine and Aquinas, who used right authority to distinguish war as a lawful activity from criminal violence, much recent work in just war theory rejects any authorization condition for war. It has become common for just war theorists to hold that while authorization may increase the likelihood that war will satisfy the *other jus ad bellum* criteria and thus can be valuable, authorization is not itself morally necessary. On this view, if a war meets the other conditions, including just cause, last resort, and proportionality, it is justified regardless of who, if anyone, authorizes it (Steinhoff 2009; 2019; Fabre 2008; Lango 2014; Lee 2012). Many just war thinkers would now agree that "the set of core just war principles does not contain a legal authority principle" (Lango 2014: 194) or any other procedural requirement.

Whether legitimate authorization is required would seem to be a fundamental question for just war theory. Authorization by appropriate bodies provides a procedural test and gives institutional legitimacy to a war. Its applicability is readily apparent in recent conflicts. For example, the 2003 US invasion of Iraq was not authorized by the United Nations Security Council according to international law. NATO's 2011 intervention in Libya was authorized by the UNSC, but US involvement was not approved by Congress according to American law. According to some views of legitimate authorization, these failures would make these two wars unjust. At the same time, that these conflicts proceeded in the face of these procedural shortcomings and that many accepted them as nonetheless valid lend support to skepticism about the authorization requirement.

DOI: 10.4324/9781003105381-8

Uncertainty about whether the UNSC and US Congress were, in fact, the proper and requisite legitimating authorities in these two cases also points to what seems to be a deep ambiguity about the nature of legitimate authority. Those who would retain the authority requirement have differing views about who the proper authority is. James Turner Johnson (2015) and other proponents of unilateral sovereign application of just war thought argue that the head of the state using force is the legitimate authority (Williams and Weigel 2004; Brown 2011; Cole 2009). Other writers, from an internationalist perspective, argue that at least for humanitarian interventions and other uses of force beyond immediate state defense, the UNSC, per UN Charter law, has the right of authorization (Syse and Ingierd 2005; Reichberg 2005; Boulette 2009).[1] Some hold that non-state actors can acquire war-waging authority (Held 2005; Finlay 2010; Schwenkenbecher 2013; Gross 2015; Eckert 2020). Thus, the nature, as well as the necessity of legitimate authority, is in dispute.

I argue that, reasonably understood, legitimate authorization is a moral requirement for just recourse to war. Deontological considerations regarding the representation of agents combine with the practical benefits of effective institutional restraints on violence to indicate the moral necessity of proper authorization.[2] This involves *prima facie* a requirement of legal authorization, including external international and internal national authorization. However, the moral arguments for legitimate authority suggest that alternate authorization is possible in appropriate contexts with proper procedure.[3]

8.2 Agency, representation, and consent in the authorization of force

Deontological and pragmatic reasons combine to support the requirement of legitimate authorization and indicate the proper authority in individual cases. The primary deontological argument for the necessity of legitimate authority derives from rights and agency in the use of force. Accounts of just cause agree that violence can only be justified in order to prevent a significant rights violation. A victim can justly use force to defend his or her own basic rights. He or she can also transfer this right to others by inviting assistance. However, the victim can also choose not to defend himself or herself or request aid. In such a case, it would be unjustifiable paternalism to use force to right an injustice without the authorization of the aggrieved party. As Christopher Finlay (2010, 294) puts it, "Third parties who proceed with a violent intervention in the face of a refusal of authorization lack legitimacy in relation to the victim-beneficiary." Intervention also almost inevitably puts the victim at new risk by involving her or him in a potentially escalated and prolonged conflict. It controverts autonomy to make life and death decisions without the consent of those centrally affected. Analogously, a medical intervention without proper authorization is considered an assault, even if otherwise reasonable and well-intended. An unauthorized intervention is, thus at least *prima facie,* a rights violation.

In some circumstances, force can justifiably be used on behalf of individuals who do not explicitly request it. A party under attack may be either unaware of the danger or unable to communicate his or her desire for assistance. Most ethical analyses consider paternalistic force justifiable in rare instances (Dworkin 1971; Beauchamp and Childress 2013), generally when there is good reason to think that the beneficiary would welcome the intervention if they had the information, were reasoning well, and could express their will. Additionally, the intervening party has to be well-situated to represent the subject of assistance, with no more appropriate agent on hand. In medical ethics, incompetent patients are to be represented, whenever possible, by a designated durable power of attorney. Parents or designated guardians are understood to have the primary right of decision-making for interventions on behalf of their children.

When it comes to using force to defend a community, governments are typically understood to represent the people's interests and have the latter's protection as their primary purpose. This representative status gives the government authority to use force to protect rights that private third parties lack. Thus the police are authorized to exercise coercion and, occasionally violence, for peacekeeping, which would be considered assault if undertaken by private citizens. The argument from the purpose of the government is bolstered by the concept of the social contract, with its suggestion that citizens accept – or hypothetically would accept – the use of force by the government. The social contract also supports the duty to refrain from private uses of force, which is part of the legitimate authority principle as a necessary condition rather than a mere permission. Citizens are understood to agree to surrender most rights of enforcing justice to the government. This includes the right to take up arms to stop crimes and administer punishment. Private citizens ought to defer to authorities in these matters rather than acting as vigilantes.

The case for deference to authorities is strengthened insofar as taking up arms centrally affects others – both the targeted opponent and any by-standers. To be justified, actions and policies must be defensible to all those affected (Habermas 1990). When an action centrally affects the rights and well-being of others, especially when it involves killing, there is a particular burden of establishing the defensibility of one's action. The decision of whether to use deadly force should be made in a manner that reliably re-presents the claims of the parties involved. The approval by a representative body charged with adjudicating rights claims should make this decision unless consultation is impossible or, perhaps, that the body has unjustifiably refused to authorize force.

The norm against unauthorized force that governs domestic disputes between individuals applies even more clearly to the use of force to combat international injustice. Modern democracies are familiar with the norm that the decision to use force should be authorized by civilian government and not by the armed forces themselves, despite the latter's inherent capacity. Private citizens do not have the right to initiate international force, for example, to counter what they judge to be an imminent threat from abroad or to undertake a humanitarian intervention. Such a decision and action would place others at risk without a representative process. If a billionaire were to, unauthorized, hire a private army to fight a war with a putatively just cause, and likely effectiveness, the use of force would lack legitimacy and be presumptively wrong for this reason.

My argument here for the rationality of following legitimation procedures draws in part on the relative reliability of authorities, a benefit which I discuss below as a consequentialist argument for the requirement. The strength of the presumed social contract and its representation of agents depends in part on the mutual advantages conferred by lawful obedience, such that it would be rational for citizens to agree to defer if others do as well. For now, I note that the consequentialist and representation arguments are mutually supportive.

The representative authority for the use of force will typically be designated by law, as legislation and institutional procedures contain the most explicit, effective, and widely recognized norms regarding rights in the use of force. The general duty to obey the law, especially regarding force, stems from both fidelity to one's promises and fairness in observing rules that one expects others to follow. The duty to obey the law is not intrinsic but instead derives from other moral values, including reciprocity, security, and fidelity. Most accounts agree that although not an absolute duty, we are generally obligated to follow the law (Smith 1996; Wellman 1996). Unless there is an emergency with imminent danger, only public authorities have the right to use force, and private parties should defer. Unlawful, vigilante violence is generally unjustified. I will take up exceptions to the principle of deference to law and elaborate on what this means for the legitimate authority requirement below.

The deontological arguments I have presented for a legitimate authorization requirement are similar to those found in Aquinas. Greg Reichberg (2012) describes these as first, the government's responsibility through its role of securing the common good, and second the principle of "no higher redress." I take these arguments to refer to the importance of decisions being made by agents who represent the rights and welfare of those concerned according to generally accepted procedures. The idea that war is valid only so long as there is no higher authority to which to appeal might be thought to be a corollary of the last resort principle (prohibiting waging war if there is a less destructive alternative). Like any additional hurdle, the requirement of authorization can lead to the avoidance of war. Reasoning in this way, Steinhoff argues that in the case of a civilian's duty to defer to the police, "'Authority' seems to have nothing to do with it: necessity is doing all the work here" (2019, p. 326). Steinhoff professes the intuition that if a civilian could neutralize a threat to himself or herself as or more easily than a police officer, "the police officer would be legally required to leave the defense to the private citizen" (326). I think this is legally and ethically mistaken: the police do not have to defer to a private actor; the opposite is generally the case. Steinhoff's focus on self-defense is also misleading as an analogy for war, which involves fighting for the defense of others, i.e. the community. Whereas an individual has a presumptive right to defend themselves, as I discussed above, this same right does not extend to the defense of others. Designated authorities have a right (and a duty) to act to defend others that takes precedence over private actors. The police do not have to defer to competent vigilantes. The vigilante is only justified in taking action if the police cannot do the job on their own or request assistance. It is authority (through the social contract, representative adjudication, and the value of shared norms), not last resort or necessity, that gives the private citizen a particular obligation to defer to the officer rather than vice versa.

The requirement of authorization at the highest level of representation can be understood to point beyond national authorization to international authorization, as I argue in Section 8.4. I would note that while I assert that procedural authorization is *necessary* for just recourse, it is not *sufficient*. Having the right body make the decision does not make the decision right. State (or UNSC, etc.) authorized recourse to force will only be *just* if the other *ad bellum* criteria (e.g. just cause, last resort, and proportionality) are also met. If the other criteria are not met, the authorized intervention will be *legitimate* but not *just*. That is, legitimacy is necessary but not sufficient for just recourse.[4]

8.3 Consequentialist considerations supporting the legitimate authority requirement

The transfer of rights to authorities is supported and supplemented by several consequentialist considerations regarding the recourse to war. First, the establishment and maintenance of procedures whereby only particular parties can authorize force are generally beneficial. A monopoly on war powers reduces the frequency of war, avoiding the unstable situation of every individual or state exercising its own prerogative regarding the re-course to violence. From a utilitarian perspective, the value of retaining institutions can override one's own consideration of what would otherwise be right to do in a particular situation.

As Walzer notes in the context of discussing *in bello* rules (1977, 133), most any accepted limit on the use of force would be utilitarian, but this doesn't pick out why a particular set of rules should be recognized. The general benefits of rules must be combined with the moral force of rights transfer through representation, the demand to make life and death deci-sions justly and reliably, and the value of supporting working institutions. In terms of prudence, the most straightforward consideration is that norms that are currently generally recognized and observed, and thus effectively furthering individual rights and communal well-being, give all a *prima facie* reason to uphold and not undermine this system. This means that legal procedures constraining war ought to be followed unless there are strong overriding considerations.[5]

The utility of violence-reducing norms combines with the demands of moral universalism, the duty to act only on principles that we could reasonably expect others to agree to and observe. Insofar as we want and expect other powers to follow international law rather than engage in vigilante uses of force whenever they see fit for preventive or humanitarian purposes, we ought to observe those norms ourselves. Pragmatic norms acquire deontological force through the demand for consistency.

A second practical consideration in favor of the authority requirement is the increased reliability of decision procedures. Because war involves killing and causes large and unpredictable destruction, there is a moral presumption against taking up arms that can only be overridden by particularly good reasons, which includes epistemic confidence. This is why the legitimacy criterion is sometimes called the principle of "competent authority." While individuals can determine whether immediate personal defense is required, they are not well-positioned to evaluate whether foreign states are planning an attack or grossly violating their own citizens' rights. Even the need for national self-defense is something private citizens are apt to lack direct knowledge about.[6] In addition to just cause, individuals are ill-equipped to determine whether a considered use of force satisfies last resort, proportionality, and likely success (Finlay 2010). The interpretation of geopolitical events requires the integration of broad data by a central body. A governing authority, with its intelligence gathering and data analysis resources and its professional and elected experts, is comparatively capable of making valid decisions regarding the use of force. Procedural checks and balances and accountability to the public, present to varying degrees in state decision-making, add quality control. These competence considerations add to the utilitarian basis for requiring that wars be sanctioned by an authority.[7]

8.4 The presumptive necessity of international authorization

Of course, states themselves can be mistaken in their war-making decisions, particularly regarding their anticipation of gathering threats or humanitarian shortcomings in other states. Not only the competence consideration, but also the considerations of contractual obligations, representation, and the effective restraint of force discussed above all point to the legitimate authority of the UNSC as the decision-making body for uses of force that are not immediately defensive. States have accepted the framework of the UN Charter and thus have a more explicit contractual obligation to observe its norms than citizens typically have to their own states. If citizens have transferred the right to authorize force to their governments, these governments, in turn, have passed that authority – at least for non-immediately defensive uses of force – to the Security Council.

In terms of representation, for international uses of force that affect people across borders as well as the stability of the international community generally, an international decision-maker is the appropriate representative body. A state can determine whether it is under immediate attack, and there is a need to respond rapidly without consulting authorities in such a situation. However, the individual state is on dubious ground deciding whether humanitarian intervention is justified on behalf of a foreign people or military developments in another state present a danger which warrants preventive or punitive attack. Unlike an individual state acting unilaterally, the UNSC could legally authorize an invasion of Iraq, Libya, Iran, or Syria to enforce international norms regarding weapons development or human rights protection (Reichberg 2005).[8] For the international use of force to combat violations of international norms, this body is the appropriate decision-maker by purposeful design and general consent.

An international body can be expected to make decisions with more reliable justice than unilateral actors. States acting unilaterally are apt to wage war for self-serving purposes, as the US did in Vietnam and Iraq, the Russians in Georgia, Iraq in Kuwait, and so forth. Although these interventions may have been sold as just to their domestic audiences, they could not secure the approval of the UNSC or the General Assembly. Requiring international authorization helps to curb the abuse of norms of humanitarian intervention, preemptive defense, or punitive war to the extent that these are justified at all. In his pragmatic defense of requiring international authorization, Stefano Recchia argues that "approval of humanitarian interventions by the UNSC, is most likely to prevent conflict spirals and accidental escalation among powerful states." (2016, 57) The requirement helps curb self-interested actions as well as competition and brinksmanship by making the sides go through an international process. Avoiding explosive regional conflicts like those between Iran, Saudi Arabia, and Israel, or India and Pakistan, and reducing tensions between powers like the US, Russia, and China is valuable for international peace. The very process of appealing to authorities may mitigate the tendency for tit-for-tat intervention spirals: "Approval by regional organizations may signal benign intentions and thus help manage international tensions" (2016, 57).

It is a classic principle of just procedures that individuals should not be judges in their own cases. The UNSC represents states from around the world and has a greater capacity for unbiased international decision-making. States are likely to have flawed information and be partial to their own interests. Appeal to the UNSC requires publicizing and deliberatively testing arguments, giving greater reliability to its sanctioned interventions than those unilaterally authorized. As Recchia argues, national governments tend to fall into "groupthink" regarding decisions to use military force, where "highly cohesive groups produce a psychological drive for consensus ... inhibit[ing] the expression of internal dissent and the consideration of alternative viewpoints – reinforcing the propensity to reach decisions prematurely, based on limited evidence and preconceived notions" (2016, 60). Groupthink seems to have characterized, for example, the US entry into the quagmires of Vietnam and Iraq (Badie 2010). Nor, as Recchia notes, is it is enough to have a multi-lateral "coalition of the willing," as some coalition whose interests overlap on intervention may always be found (56). An impartial body, or at least one representing varied interests and perspectives from around the world, especially the affected regions, is required to test the validity of an intervention.

Although in individual cases, international legal procedures may yield flawed results, the general superiority of their decision-making in tandem with the value of restrictive procedures, the authority of representative bodies, and the commitment to observe recognized principles all imply a requirement of international authorization for armed intervention. Individual states, like individual citizens, have an obligation to appeal to higher authorities, generally defer to their judgment, and refrain from international vigilantism (Rocheleau 2007; Boulette 2009).

In addition to curbing violence between potential interveners, Recchia argues that an international authorization requirement can reduce the "moral hazard" regarding humanitarian intervention in particular. In Chapter Five, I introduced Kuperman's idea that permissiveness of intervention may encourage rebel groups to escalate conflict and refuse to negotiate in the hope of drawing intervention. As Recchia points out that this is especially likely to be effective if the morally hazarding body only needs to persuade one powerful unilateral intervener. If there is a multi-lateral international test, then it is more difficult to secure intervention. Moreover, multi-lateral deliberation would be more likely to expose any inflammatory role of rebel groups. These factors should reduce perverse incentives to escalate violence (Recchia 2016, 63–6).

International law's detractors reject international institutions' capacity to justly represent local parties and pursue the common good. Robert Keohane points out that the UNSC composition is a modus operandi based on the balance of power: it "does not reflect any principled set of criteria for representation ... [and] cannot be justified on the basis of principles of either democracy or elementary fairness" (Keohane 2006). The veto possessed by five powerful states over any decision, no matter how broadly and reasonably supported, is the most extreme example of this undemocratic and unfair structure.

In support of his view that international bodies are not legitimate authorities and their approval not necessary to justify force, Davis Brown argues that UN decision-making is "thoroughly political," not driven by concern for the greater good and impartial justice. He is particularly concerned that "as the hegemon in the international state system, the United

States is particularly susceptible to unfair treatment" (2011, 137). As evidence, he cites three World Court cases involving the US as well as UNSC decisions condemning Israel as evidence of the political bias of international institutions. He infers that "the Court's [and the UNSC's] past susceptibility to geopolitical influences does not instill confidence in its ability to judge cases of aggression impartially in the future" (136–7). Because of the failure of international bodies to impartially protect the common good, Brown (143) concludes that "the traditional approach of states judging their own causes is still the best hope for the promulgation of justice, world public order, and a true peace."

While the UN and UNSC are certainly flawed, the argument from these limitations to a unilateral right of war-waging by states is a non-sequitur. The politics and bias that Keohane and Brown find fault with at the UN are all introduced by state governments. If the UN as a collective body fails to impartially pursue the common good, this is certainly true of individual states. Brown fails to consider whether states themselves are influenced by "politics" in their exercise of force. A unilateral intervener is much more limited in perspective and interests, prone to bias and groupthink, and gives no representation to any external party, including the foreign citizens centrally affected by its use of military force.

Oddly, in none of the World Court cases Brown cites to question the legitimacy of international law – that finding Iran responsible for the US hostages in Tehran, that finding the US responsible for attempting to overthrow the Nicaraguan Sandinista government, and a decision regarding US responsibility for destroying Iranian oil platforms in 1978 – does Brown argue that the decision was made wrongly. Instead, he criticizes the "anti-American tone" of the discussions (reflected in a dissenting Soviet vote in the first case and a failure to condemn actions by Nicaragua and Iran related to the US aggression.) These decisions show a quite reasonable application of international law about which there is little to complain. Indeed in both the hostages and oil platform cases, the court found in favor of the US (refusing to condemn either the US or Iran in the latter). The court's finding against the US for backing the Contras' attempt to overthrow the Sandinista government was certainly justified. Brown only presents one case in which he claims that an attack was unjustifiably condemned by the UNSC: Israel's preventive destruction of Iraq's Osirak reactor. However, that condemnation was a straightforward application of international law as Israel's strike neither responded to an imminent attack nor sought UNSC Authorization. Even the Reagan administration – hardly an anti-Israeli body – joined the condemnation.

If these are the worst cases of international decision-making, it does little to call into question its superior justice and legitimacy. I suggest that if one looks at the relatively long list of UN-authorized interventions since World War II, there is a lack of decidedly unjust wars. The glaringly unjust interventions over the past decades, such as the US invasion of Vietnam and

Iraq, Iraq's invasion of Kuwait, and the Soviet intervention in Afghanistan and the Russian incursions in Georgia and Ukraine, have lacked UN authorization.

The UNSC is most frequently criticized for its *inaction*. Yet, there are few cases in which its refusal to act has resulted in disastrous consequences. This is in part because it typically has been willing to authorize interventions in the most pressing cases – e.g. East Timor and, although insufficiently robust, Rwanda and Darfur. Moreover, in those cases in which international action was culpably lacking, there was also a lack of will on the part of powerful states to intervene, so that it was not international legal procedures that prevented effective intervention. It is not the case, as critics of internationalism frequently imply, that obstruction at the UNSC kept willing and eager humanitarian democracies from intervening sufficiently to stop the genocides in Rwanda and Darfur.

Historically, the greater danger has been illegal unilateral intervention rather than legalistic restrictions on force. The unpredictable destructiveness of war, and the necessity of justifying it to those affected, support the authority of an international legal body even when this body is flawed. International norms regulating the use of force benefit all the world's states, including the US and Israel. Conforming to reciprocal agreements restricting the first use of force is in the individual and collective interest. International law already permits defense against immediate attacks, so there is no question of sacrificing a state right of self-defense, as detractors of international authority sometimes suggest.

Some argue that international law and the UNSC lack legitimacy because they do not have consistent enforcement power (Eckert 2020). In response, authority does not stem solely from coercive force. To gain legitimacy, an authorizing body does not itself have to carry out the action: for example, the US Congress has the authority to declare war though it isn't the party that will fight. Its legitimacy is not based upon its ability to forcefully compel the military to act or refrain from action. There is an ethical obligation to submit to rules and institutions to which one has agreed, which represent the parties affected by one's belligerent plans, and which helps to ensure a good decision. This holds even if other parties do not consistently comply.

However, there is also reason to think that international law, and UNSC decisions, in particular, do exercise power, and this helps undergird their legitimacy. Recchia (2016) cites research finding that international actors are very concerned to secure international approval and avoid censure. Governments in democratic states rely on international approval to be able to defend the justice of their wars to a sometimes skeptical domestic audience. Even non-democracies are concerned to be in good international standing. The flouting of international norms – although it usually will not incur UNSC-endorsed military punishment – risks retaliation in the form of economic sanctions or the refusal of the benefits of economic cooperation. International condemnation or "outcasting" has significant repercussions for states and motivates compliance with norms (Hathaway and Shapiro, 2014). In general, most states are not fighting for their survival and thus will only consider military intervention if it is at a low cost. This, in turn, means that "[m]ultilateral (international) institutions have the ability to constrain powerful states by threatening to delegitimize, and thus increase the cost, of interventions that are not clearly in self-defense and are carried out without those institutions' approval" (Recchia 2016, 60).

It is important to note that the authority of a body can be recognized without holding that its decisions are perfect or that the duty to obey its rulings is absolute. UN norms could be made more just and effective, for example, by getting rid of the unfair and overly restrictive veto and holding states accountable for flawed decisions (Buchanan and Keohane 2004). In the meantime, the proper response to a flawed institution is not to deny its authority – and especially not to assert the adequacy of more partial and flawed states across the world. Rather the developing institution should be supported to increase its effectiveness (Boulette 2009). International authorization, with its indication of widespread support, also makes it more likely that the intervention will succeed, adding to its pragmatic basis and thus legitimacy.

One might wonder why we should rely on the competence of existing flawed institutions. Why not instead require the approval of something like an "impartial international court" stocked with just war scholars as envisioned by Jeff McMahan (2009, 153) or another body with a history of ethical decision-making superior to that of the UNSC? Writers such as Allen Buchanan (2003; 2010) and John Davenport (2011) have argued that considerations of legitimacy and human rights require that we establish a federation of democratic states to authorize humanitarian interventions. In response, for competent judgments about warfare to effectively constrain war, they must be made by political bodies with existing institutional capacity and public recognition; that is, those who can actually either send or refuse to send forces to war. The stability of existing institutions, with their functional value in securing the peace, gives reason to follow flawed procedures rather than some more ideal authority that is not legitimized by effective recognition. None of this implies that we could not change UN rules or form new institutions. Yet, the value of existing institutions implies caution about implementing procedures bypassing and undermining those operating currently. Davenport and Buchanan are also too sanguine about the potential relative effectiveness of a democratic federation in upholding rights, given the historic imperialism and rights abuses on the part of these democratic states. The voices of third-world countries must be represented in decisions about international uses of force if the decisions are to legitimately represent those centrally affected. To re-emphasize my main point, the mere potential of a more just system does not take away from the legitimacy of the currently recognized and operational legal system.

8.5 The internal legitimacy requirement

Resistance to the concept of international authority is motivated in part by a worry that deference to the UN would mean the loss of national control over armed forces. Under current law, this fear is based on a misconception, as UNSC authorization does not *require* states to send forces to support an intervention; it only legitimates their doing so. States retain the prerogative whether to risk their own citizens' lives and resources on each UNSC-sanctioned mission. This legal limit has a principled basis. While the United Nations is the appropriate body to determine the validity of international intervention according to global norms, the state government remains obliged to represent the rights and interests of its own citizens and armed forces. While national interests and security cannot justify overriding international norms prohibiting intervention, they can give reason to refuse to participate in an otherwise just and legitimate intervention. As Walzer argues, states have a right to remain neutral because by entering a war, they take on "risks that may be acceptable or not" and "condemn an indefinite number of [their] citizens to death." The state is charged with protecting its citizens' welfare, including those in the military, and thus has an obligation to responsibly oversee their deployment. A state's moral goals might justify or even morally obligate it to become involved in an emergency, but these do not waive the general right of self-determination.[9]

This implies that interventions should have what Buchanan (2010) calls "internal legitimacy" – support from the intervening political community – as well as the "external legitimacy" of the intervened-in state or world community. War should be legitimized to all of these stakeholders. Regarding the 2011 US intervention in Libya, though it was externally (i.e. internationally) legitimate, its internal legitimacy was questionable because of the lack of US congressional authorization.[10] To be legitimate, recourse to war must be approved by appropriate representatives of all those affected, including international and national representatives.

8.6 Military effectiveness, restraint, and authority

A final set of utility considerations also speaks in favor of the authorization requirement. Having parties to a conflict governed by authorities helps to ethically constrain the conduct of warfare and facilitate the negotiation of peace. An established chain of command with a regime of accountability is required if forces are to fight according to the rules of war. Governance by a representative and legally recognized authority is important for enabling the effective pursuit of military goals, including the achievement of the just cause and a justly concluded peace. Political leadership is conducive to, if not absolutely necessary for, the diplomacy that could reach an agreement ending the war. Effective institutional restraint is central to Janna Thompson's (2005) conception of legitimate authority: "first of all, it must be accountable for the violence of its members; it must be able and willing to enforce the restrictions of just war theory, to negotiate a peace and to keep it." Whether or not a commitment to *jus in bello* is necessary for *jus ad bellum* as Thompson, among others, suggests, an ability to organize and restrain the use of force is centrally relevant to the prospect of the proportionality and likely success of the conflict, necessary criteria for *jus ad bellum*.

These benefits from institutional control over the waging of war point most directly to the *internal* legitimacy requirement. However, international recognition and authorization of a war-waging party also provide reassurance of its ability to fight ethically and achieve a just peace. International legitimation indicates that a war cause has wide support and encourages other parties to join the cause. This makes the authorized intervention more likely to achieve its goals, including a just peace with effective post-war reconstruction. Effective reconstruction also requires an ongoing commitment, with public support in the intervening country. Citizens are more likely to be willing to endure the costs (which will, in turn, be proportionately reduced) if there is international burden sharing. "Such burden sharing is significantly more likely to materialize if the original intervention was approved by appropriate international bodies; indeed, multilateral approval may be essential for longer-term burden sharing (and thus ultimately for successful peacebuilding)" (Recchia 2016, 66). Similarly, by lending legitimacy, international approval reduces moral objections that can erode national commitment to the cause. These considerations about the requirements for a successful, proportionate mission provide further support for the necessity of legitimate authorization.

8.7 Defending a legitimate authorization requirement

Anne Schwenkenbecher has argued that the pragmatic considerations regarding the advantages of legitimate authority that I have outlined are "redundant," as seeking authorization on these grounds is already implied by the other just war criteria, especially proportionality and likely success. Thus she concludes that they fail to lend support to legitimate authorization as an independent criterion. For Schwenkenbecher (2013), it is only the value of "representation of a people" (akin to my first rationale above) that provides reason to require legitimate authority.[11] For similar reasons, Steinhoff, who rejects duties to defer to authorities based upon consent and representation, argues that to the extent that authorization is necessary, it is entirely derivative of the other criteria and thus "redundant" and "superfluous" (2019, 315).

In my view, that authorization procedures contribute to the satisfaction of other just war principles does not entail that the irrelevance of these benefits to the justification of a distinct authorization requirement. On the contrary, competent procedures are valued in part because of their rule utilitarian tendency to produce good results. The value of authorization procedures, especially legally recognized ones, lies in part in their helping to ensure just decisions. Conviction by a court is a moral requirement before punishment in part because the procedure makes undeserved punishment less likely and lawfulness prevents a cycle of blood lust, revenge, and other biased and unjustified attacks. The constellation of benefits that attach to following rules provides a strong *prima facie* case for following the law. Adhering to institutional processes is required for adequate consideration of the competing claims of the involved parties as well as to fulfill deontological contractual obligations.

The duty to obey the law is not absolute and depends on the norm under consideration and the ethical consequences of following or not following it in a given situation. However, at least for life and death decisions, rationality and respect require an appropriate procedure. The appropriate authority will *prima facie* be a legally recognized one. The nature of this institutional body (whether the UNSC or a domestic court) is historically contingent, and the feasibility of consulting it in every case is circumstantially contingent, but to the extent that they are available and there is time to employ them, applying them is morally necessary and not merely recommended.

There are other challenging objections to my case for the necessity of legitimate authorization. The extent of individual agreement to transfer rights to and obey authorities is notoriously dubious. The terms of the social contract consented to tacitly or hypothetically are far from clear. The case for a transfer of rights or an obligation of deference is further weakened when a government fails to represent its people or is otherwise unjust in its use of force. In any case, it is clear that contractual promises and participation in a legal system do not bind parties absolutely. It would be an odd moral view that held that promise-keeping and rule-following are more important than every other norm or value, up to and including preventing genocide. While it is generally valuable to follow legal procedures, it is undeniable that there are exceptions in which the authorities get it wrong, and the duty to fight injustice outweighs the duty to obey the law. A party might intend to break the law in just those cases in which it would have others do so as well so that its illegal action satisfies moral universalizability. For example, if the US had been inclined to use its forces to stop the genocides in Rwanda, it seems that it would have been justified in doing so, lack of UNSC authorization notwithstanding. Even a private army might be justified in acting if necessary and feasible in such a case. Such considerations imply that while authorization may help to confirm justice, it is not strictly required for a war to be just.[12]

In this vein, Ned Dobos (2010, 511) argues that "states and coalitions must seek a mandate from the U.N. ... but they need not refrain from unilateral action if a mandate is not forthcoming." A requirement of *consultation* would allow one to draw on the benefits of deliberative procedures and show respect for institutions while avoiding rigid, legalistic deference. The case for consulting but not deferring to authorities appears strong.

In defense of the necessity of legitimate authority, I would make two points. First, the above considerations of representation, contractual obligations, stability, competence, and effective restraint, do not just give reason to *consult* authorities but also give *prima facie* reason to *defer* to their judgment. Substituting private judgment causes mistakes and is destabilizing, such that it should not be done lightly. Nonetheless, the obligation to defer is not absolute. If an authority fails to undertake its deliberative obligations and is clearly and disastrously mistaken, then one may bypass that authority. However, and this is the second key to my argument, armed disobedience is only valid if it itself has an appropriate alternative authority, following representative and effective procedures. The same rationales that support *prima facie* deference to authority imply that the overriding of primary authorities should itself be subjected to proper procedures. If the UNSC fails to exercise its authority, the next level of representative and recognized institutional authority is a regional security alliance that contains the states involved in the conflict. Similarly, if the regional alliance fails or refuses to adequately consider a serious threat to human rights after being petitioned, an individual state could gain authority to act independently. In fact, this sort of procedure, in which default authorities can be overridden in rare circumstances, is written into the new "Responsibility to Protect" norm:

> If the Security Council rejects a proposal or fails to deal with it in a reasonable time, alternative options are: consideration of the matter by the General Assembly in Emergency Special Session under the 'Uniting for Peace' procedure; and action within area of jurisdiction by regional or sub-regional organizations under Chapter VIII of the Charter, subject to their seeking subsequent authorization from the Security Council. (International Commission on Intervention and State Sovereignty 2001)

Groups such as NATO, the African Union, or the Organization of American States can provide multilateral accountability, checks on power, epistemic testing, and mutual support as well as the UNSC (Recchia). I would emphasize the importance of following procedures, whereby the regional body is only used if necessary following UNSC intransigence and if geographically appropriate. If a party skips the UN and goes straight to a regional body, this both undermines the authority of the UN and sets up a system in which regional groups may act in a lawless and self-serving manner. I will add that it would be legitimate, if not legal, for a state, first, or sub-state actor, second, to intervene in extreme cases if there is a failure to act at all of the above levels. Such bodies provide a backup procedural authorization, such that necessary extra-legal uses of force can occur but in a process that involves rational testing and accountability constraints.

My proposed conception of right authority might be charged with emptiness. If a refusal of authorization of war at one level gives the parties at a subsidiary level the right to self-authorize, then it might seem that right authority will always automatically be met, for any party that fights will either have had its war authorized by others above it or be taking it upon itself to authorize its own war in the face of their inaction. Steven Lee (2012, 83) notes that the authority requirement risks being an empty tautology since "almost all war would satisfy the criterion, given that the de facto rulers of organizations capable of using force act under the rules of that organization."[13]

In response, my conception of legitimate authority has substance insofar as it gives precedence to the highest level of representation and legal agreement, principally international authority. I emphasize that a subsidiary agent may only justifiably circumvent higher authorities if it has good reason to think that they have failed in exercising their authority. Mere disagreement with their ruling is not enough. If serious debate leads to a refusal to authorize intervention, the result must be recognized as legitimate and binding.

However, if a decision process is irrational, such as by ignoring evidence of human rights violations, then a party at the next institutional level of representation – ideally itself legally recognized – can legitimately authorize the war without breaking its moral contract or necessarily undermining the potential of the higher authority to make decisions in the future. Principled action after petitioning and spelling out one's objections to the higher authority could be viewed as a form of "armed disobedience" and could encourage the reform of international institutions at the same time that it accords them respect (Hoag 2007).[14] Representation and effectiveness are preserved by moving to an alternative authority. The process of devolution should itself be generally acceptable. The UN report concluding that the NATO intervention in Kosovo without UNSC authorization was nonetheless legitimate could be understood to follow this model, as the exception was principled and involved authorization by an arguably relevant regional body (The Independent International Commission on Kosovo 2000).[15] By contrast, the invasion of Iraq was a clear case of a lack of legitimate authority. The UNSC's refusal to endorse the 2003 US invasion was hardly irrational, and in bypassing that body, the US and its allies did not secure the authorization of a relevant regional group – the Arab League or even NATO. When higher authorities have not been consulted and have not abdicated their responsibilities, lower level authorization, especially unilateral authorization, is inappropriate. The requirement stands that decisions to wage war will be unjust if they are not made by an appropriate body.

One might object that my view makes legitimate authority ambiguous, dependent on interpretations of what is required by law and when the law should be bypassed. The continuing dispute about the legitimacy of the Kosovo intervention is a case in point. In response, the existence of difficult cases and the need for interpretation does not make the authority principle empty. It directs us to matters which can be rationally debated and are not merely arbitrary. Difficult, borderline cases do not make a requirement empty or unnecessary. If a modicum of vagueness were a refutation, we would have to dismiss all the just war criteria. As with the other just war conditions, there are clear cases of failure of legitimate authorization, and this condemns those wars as unjust.

8.8 Statism and legitimate authority

While my analysis of legitimate authority is primarily international, it retains a role for state governments as legitimate authorities for emergency state defense and last resort in cases of law enforcement in which international authorities fail. This raises the question of the authority of tyrannical, seemingly illegitimate, governments. Against the recent tendency to equate legitimacy with democratic representation, I defend the legalist view that any state that controls territory and military forces should be recognized as a legitimate decision-maker for national defense. Legitimacy is acquired by any party which effectively exercises power and is recognized as authoritative. Dictatorial governments, although significantly unjust and deserving of the overthrow, can still validly exercise force for legitimate purposes. Such bodies are able to and frequently do use force in a way that represents the people governed and furthers their rights and well-being. Stalin was justified in leading the Soviet Union in its defense against the Nazis and Saddam Hussein the Iraqis against the American invasion. Moreover, if the international community and regional authorities fail to intervene to stop gross human rights violations, an authoritarian government can legitimately do so. Tanzania's intervention to overthrow Idi Amin in Uganda and Vietnam's stopping of the killing fields in Cambodia were candidates for valid self-authorization but were of dubious legitimacy because they did not appeal to the UNSC first. This unilateralism paralleled their self-interested intentions and a lack of focus on human rights protection with mixed results. Recchia takes this to confirm, or at least conform to, his argument for the necessity of multilateral intervention (2016, 69–70). I would argue instead that it shows that unilateral actors must first seek international approval and only intervene unilaterally if their actions are unreasonably rejected. In principle, unilateral authorization can be beneficial and still respect institutions and multilateral interests. Nonetheless, the primacy of international authority means that most interventions undertaken unilaterally, whether by liberal or illiberal states, will be illegitimate.

In any case, the blanket rejection of undemocratic states as illegitimate does not hold. As I argued in Chapter Four, this would make the governments of half the world and most of the Middle East, Africa, and Asia illegitimate, such that they cannot even authorize defensive wars. Moreover, there were no legitimate uses of power until roughly 200 years ago with the first representative democracies. This is highly implausible. Undemocratic authorities have used force to preserve the peace and further human welfare and, ultimately, rights. Justice entails that democracy should be encouraged and its absence criticized. However, in the meantime in the world, as it is non-democracies must be recognized and as legitimate decision-makers regarding the use of their military forces within the framework of international law.

8.9 Non-state actors and legitimate authority

A defense of legitimate authority must also address the question of insurgents and other sub-state armed groups. A frequent criticism of the authority criterion, made for example by both Fabre (2008) and Steinhoff (2009), is that it implausibly prohibits *all* revolutionary violence. On any reasonable view, non-state actors are sometimes justified in taking up arms, typically against their own unjust government (Gross 2015; Eckert 2020), but also to defend against foreign aggression or conduct a humanitarian intervention where other authorities have failed. Fabre and Steinhoff both argue that even a single individual could justly decide to wage war if the other *ad bellum* criteria, including last resort, are satisfied. Steinhoff takes this as a decisive counterexample to the legitimate authority criterion: "Can individuals as individuals … justifiably wage war? If the answer to this question is 'yes,' then legitimate authority is not a valid part of just war theory" (Steinhoff 2019).

In response, it is an extension of my view that if a state government fails to uphold basic rights, fulfilling its obligations, a sub-state group could assume authority for exercising force on behalf of the people. There is a difficult question about how a particular non-state group can gain authority. Not every armed group within a rights-abusing or non-rights upholding state is itself a valid candidate for leading an armed response. The group must have a good case that it represents the people on whose behalf it struggles, and that there is no more representative, authorized body which it is ignoring. International law already recognizes insurgent forces that control territory or are engaged in a struggle for national liberation from colonial domination as lawful belligerents. As other authors have pointed out, there are signals of popular support short of controlling territory or holding elections, including political polls, citizen participation in mass demonstrations and strikes, and material support of the rebel cause. I second Virginia Held's (2005, 188) conclusion that "the requirement of legitimate authority should not be thought impossible to meet for non-state groups using violence." Groups such as the PLO, IRA, FLN, and ANC exhibited representative legitimacy without holding territory. Gross (2015) and Eckert (2020) also contribute to the articulation of legitimacy criteria for non-state actors. Such theories are consistent with my proposal.[16,17]

I concur with Steinhoff and Fabre that, in principle, in an unusual case, even an individual might legitimately wage war. However, in the freakish instance in which it were justified, it would be compatible with the authority requirement. For the individual's recourse to arms to be a "war," by definition, it must be on behalf of a wider communal political cause. For the individual to legitimately decide unilaterally to take up arms on behalf of the community, every broader social or political group (not just the government) would have to have proved incapable of or unwilling to consider the cause. If his or her community rejects the conflict after reasonable deliberation, he or she is bound to refrain from taking up arms on their behalf against their will. In the unlikely scenario that every other responsible decision-maker is captured, brainwashed, or otherwise unable to communicate, then and only then could our heroic individual make the unilateral decision to go to war.

My conception of legitimate authority can be usefully contrasted with Nicholas Fotion's. He argues that there is an "asymmetry" in the legitimate authority criterion such that it only applies to regular wars and national groups, but not irregular fighters in irregular wars.

> All just war theory I [for irregular wars] requires is that they use these authorities when they are supposed to do so, but typically non-nation groups do not have legitimate authorities in the midst … Since they are unable to satisfy this principle, just war theory I [irregular] does not ask them to. (Fotion 2007, 121)

There are at least a couple of problems with this. The idea that there are two different sets of just war principles seems to arbitrarily and overly rigidly bifurcate morality. Moreover, Fotion's view makes the legitimate authority criterion empty since it says that war must be authorized by institutions unless it is not so authorized (as an irregular war). My view is substantive in saying that sub-state belligerents must be legitimately authorized and suggesting a way to adjudicate whether this has occurred.

By demanding explicit evidence of representativeness of non-state groups but not government forces, my view is open to a charge of inconsistency. Why don't the same substantive requirements that apply to extra-legal insurgents apply to governing authorities? In reply, by gaining recognition and exercising effective control, a state actor is *prima facie* an authority representing interests in the national and global communities, even if not necessarily a just one. A group lacking power has to provide other evidence of its prospect for acting on the people's behalf and not simply as criminals exercising violence without political representation (Finlay 2010). Governments *should* take their own population's views into account in deciding whether a war on their behalf is justified. However, as I argued earlier, the social contract, along with overall effectiveness and stability, give institutions the authority to make decisions to use force on their people's behalf in accordance with their judgment. For either national defense or humanitarian intervention, it is a mistake to conceive of legitimacy as the echoing of popular opinion. I disagree, for example, with Schwenkenbecher's (2013) suggestion that Spain's contribution of forces to the Iraq War was illegitimate because it was opposed by the overwhelming majority of the Spanish people. Assuming that the government decision was within its proscribed authority, in accord with legal procedures, the government acted as the people's legitimate representative even if the people disagreed at the time. This is not, of course, to say that people may not criticize their officials and seek to replace them for making unjust decisions.

8.10 Reply to final superfluity objection and conclusion

In explaining who the legitimate authority is and why, particularly for non-state groups, I have referred to matters including the group's cause and likely effectiveness. Since these matters are addressed in other *jus ad bellum* criteria, one might wonder whether the legitimacy criterion is derivative after all and thus superfluous. In reply, my analysis of legitimate authority is irreducible because it requires procedures that represent moral agency and provide effective institutional restraints. The satisfaction of the other just war criteria is not sufficient for legitimacy. A war could meet all the other *ad bellum* criteria but be illegitimate due to a flawed authorization procedure. On the other hand, a legitimate authority can make an unjust decision, further indicating the independence of the criterion.

For example, the US entry into the Spanish-American war was based in large part on a mistaken judgment that Spain had committed aggression by sinking its ship and that conflict lacked just cause. However, given the lack of international authority at the time, the US government was the appropriate decision-making body. Similarly, even if, as Walzer (1977, 78–80) argues, the British were unjustified in launching the War over the Spanish Succession against Louis XIV's France, the British government was the legitimate decision-maker regarding its protection.[18] More controversially in the US Civil War, even if, as I assume, the Confederacy lacked just cause, its representative and deliberative procedures gave it political legitimacy to make this sort of decision. The Confederate government formed by the Southern states was the appropriate body to make a decision regarding the use of force to secure separation from the Union. It uniquely represented the aggrieved party.[19] Although it may sound odd to some to say that a group can "legitimately" make unjust decisions, this is the implication of the intelligible distinction between justice and legitimacy. My view of right authority as proper procedure retains this distinction that opponents of legitimate authority would erase.

In other cases, a war that would otherwise be just can be immoral because of its failure of legitimate authority. The arguable lack of proper authorization cast the Kosovo War into doubt, and its clear violation condemns the Iraq War and Israel's Osirak attack. Even an effective humanitarian intervention with an indisputably just cause will be unjustified, despite its satisfaction with other *ad bellum* principles, if it fails to seek and secure UNSC authorization or receive legitimation by a representative body at a subsidiary level. Tanzania's Uganda and Vietnam's Cambodia interventions arguably were unjust for this reason.

I have defended the necessity of a legitimate authorization process for the justification of war. Deontological and consequentialist rationales combine to support deference to decision-making procedures. Legitimate authority derives from representation, rationality, and the general epistemic and pragmatic value of institutional norms. These retain force independently of whether one judges their decisions to be correct in an individual instance. This serves to generally condemn unilateral state action in contravention to international law as well as force by private, sub-state groups. Among responses to objections to the necessity of authorization, I emphasized that in particular situations, authority can be taken up at a lower level, even bypassing the law. I have thus sought to clarify the nature of legitimate authority. Properly understood, legitimate authorization is a necessary and substantive criterion for *jus ad bellum*.

Notes

1 Some, like Matlary (2004), have sketched a middle view, in which either the UNSC, states, or other parties can give legitimacy depending on the context. I flesh out a similar position below.
2 My division into deontological and consequentialist rationales has some similarities to Syse and Ingierd's (2005) distinction between "procedural" and "moral" reasons for the authorization requirement. However, I consider procedural concerns to have moral import. I hope to clarify the difference and interrelation between pragmatic and deontological rationales.
3 The question of legitimate authorization should be distinguished from that of which forces should intervene, which is the guiding question for, for example, Pattison (2008).
4 Admittedly, ordinary language frequently uses "just" and "legitimate" interchangeably, conflating this distinction between right procedure and the overall correctness of an action. I take it that my distinction is intelligible, roughly aligns with usage by other political philosophers, and has the moral implications defended here.
5 Indeed Walzer himself is too dismissive of the significance of positive norms as a result of his emphasis on their contingency on other moral arguments.
6 As I discuss below, in addition to epistemic concerns, individual decisions to resort to violence are unlikely to be effective in achieving just results.
7 I discuss the problem of the legitimate authority of unrepresentative and unjust governments in Section 8.8.
8 The UN Charter only allows for intervention when authorized by the UNSC (United Nations, 1945, Chapter VII). This principle is reiterated in the recent "Responsibility to Protect" (International Commission on Intervention and State Sovereignty, 2001).

9 The logic of the right not to be forced into war ultimately implies that individual combatants should have a right to opt out of going to war. Although arguably those who have enlisted have agreed to go to wars as ordered, I would support a right of "selective conscientious objection," in which combatants may opt out of wars that they sincerely believe are unjust.

10 President Obama argued that the congressional control over war authorization did not apply because the conflict involved relatively minor risks and costs to the US and its fighters (Frowe 2016, 235). One might argue that unilateral presidential action has become a new legal norm or that Congress' failure to act justified emergency presidential action. I will not attempt to adjudicate these legal questions here.

11 Interestingly, Schwenkenbecher's position is diametrically opposed to that of Recchia. For Recchia, it is entirely the practical benefits that imply a legitimate authority requirement. In my view, it is a combination of the two. The social contract and representation through legal procedures give a *prima facie* indication of authority, and the rule utilitarian benefits give additional duty to observe those procedural norms.

12 Recchia (2016) argues that the benefits of multilateral procedures imply the injustice of any unilateral intervention. I think this is too strict. A norm of consultation of international authorities, general deference, and appeal to the next most representative body can allow us to both avoid dangerous unilateralism and permit exceptional cases in emergencies.

13 In a similar vein, Jonathan Parry (2015) takes the legitimate authority requirement to be definitive of war, such that it is necessarily met in any genuine war. Legitimate authority would then be an *in bello* requirement for troops to be legitimate combatants as opposed to criminals. My view is that there might be wars, communally organized, which do not have legitimate, proper procedural authorization. Combatants in such a conflict would not necessarily be criminal even though the leadership's decision to enter the war was illegitimate and, thus, unjust.

14 I am influenced by Bob Hoag's (2007) defense of "violent international disobedience." However, my view is more restrictive of unilateralism, consistent with the authorization requirement, by emphasizing that parties should (a) first appeal to the UNSC, (b) defer to it if it isn't clearly mistaken, and (c) appeal to the next level of representative authority before acting unilaterally.

15 I question whether the prospect of a Russian veto was based on such irrational and self-interested motives that circumventing the UNSC was warranted in that case. The other just war criteria were not so clearly met that the Council had lost its authority.

16 A complicating factor is that insurgent groups almost inevitably begin with minor levels of support and only gather more public approval following their engagement with state or occupying forces. Although I cannot develop the details of this proposal here, I would begin by suggesting that the early rebellion can be legitimated by taking up arms for a just cause, in politically repressive circumstances, and with an effort to represent the interests of the people, and that in later stages it can and should have wider levels of support in order to continue as legitimate. Some non-state actors, like Al Qaeda and Isis, remain unrepresentative of the people on behalf of whose interests they claim to act.

17 During a civil war, such that both parties possess power over territory and represent people therein, each side may possess legitimate authority.

18 This is another illustration of the advantage of having an international authority rather than unilateral decision-making.

19 Of course, the Confederacy failed to justly represent and respect the rights of much of its population – particularly African American slaves. I argued earlier that authority cannot be limited to only just bodies. At the same time, even democratically legitimate states can make unjust decisions.

References

Badie, Dina. 2010. "Groupthink, Iraq, and the War on Terror." *Foreign Policy Analysis* 6 (4): 277–96.

Beauchamp, Tom, and James Childress. 2013. *Principles of Biomedical Ethics*, 7th edition. New York: Oxford University Press.

Boulette, Michael. 2009. "Questioning Authority: Just War Theory, Sovereign Authority and the United Nations." *Vera Lex* 10: 145–68.

Brown, Davis. 2011. "Judging the Judges: Evaluating Challenges to Proper Authority in Just War Theory." *Journal of Military Ethics* 10 (3): 133–46.

Buchanan, Allen. 2003. *Justice, Legitimacy, and Self-Determination*. Oxford: Oxford University Press.

Buchanan, Allen. 2010. "Institutionalizing the Just War." In *Human Rights, Legitimacy, and the Use of Force*, edited by Allen Buchanan, 250–79. Oxford: Oxford University Press.

Buchanan, Allen, and Robert Keohane. 2004. "The Preventive Use of Force: a Cosmopolitan Institutional Proposal." *Ethics & International Affairs* 18 (1): 1–22.

Cole, Darrell. 2009. "Thomas Aquinas on Virtuous Warfare." *Journal of Religious Ethics* 27 (1): 57–80.

Davenport, John J. 2011. "Just War Theory, Humanitarian Intervention, and the Need for a Democratic Federation." *Journal of Religious Ethics* 39 (3): 493–555. 10.1111/j.1467-9795.2011.00491.x

Dobos, Ned. 2010. "Is U.N. Security Council Authorization for Armed Humanitarian Intervention Morally Necessary?" *Philosophia* 38: 499–515. 10.1 007/s11406-009-9233-1

Dworkin, Gerald. 1971. "Paternalism." In *Morality and the Law*, edited by Richard Wasserstrom, 107–26. Belmont, CA: Wadsworth Publishing.

Eckert, Amy. 2020. "The Changing Nature of Legitimate Authority." *Journal of Military Ethics* 19 (2): 84–98.

Fabre, Cecile. 2008. "Cosmopolitanism, Just War Theory, and Legitimate Authority." *International Affairs* 84 (5): 963–76.

Finlay, Christopher. 2010. "Legitimacy and Non-State Political Violence." *Journal of Political Philosophy* 18 (3): 287–312.

Fotion, Nicholas. 2007. *War and Ethics: a New Just War Theory*. New York: Continuum.

Frowe, Helen. 2016. *The Ethics of War and Peace*, 2nd edition. New York: Routledge.

Gross, Michael. 2015. *The Ethics of Insurgency*. Cambridge University Press.

Habermas, Jurgen. 1990. *Moral Consciousness and Communicative Action*. Cambridge, MA: MIT Press.

Hathaway, Oona, and Scott Shapiro. 2014. "Outcasting: Enforcement in Domestic and International Law." In *Philosophy of Law*, 9th edition, edited by Joel Feinberg, Jules Coleman, and Christopher Kutz, 299–314. Boston: Wadsworth.

Held, Virginia. 2005. "Legitimate Authority in Non-state Groups Using Violence." *Journal of Social Philosophy* 36 (2): 175–93.

Hoag, Robert. 2007. "Violent Civil Disobedience: Defending Human Rights, Rethinking Just War." In *Rethinking the Just War Tradition*, edited by Michael Brough, John Lango, and Harry van der Linden. Albany: SUNY Press.

International Commission on Intervention and State Sovereignty. 2001. "The Responsibility to Protect." Accessed July 12, 2015. http://responsibilitytoprotect.org/ICISS%20Report.pdf

The Independent International Commission on Kosovo. 2000. *"The Kosovo Report."* Oxford: Oxford University Press.

Johnson, James T. 2015. "Just War, As it Was and Is." *First Things* January, 2015. Accessed June 30, 2016. http://www.firstthings.com/article/2005/01/just-waras-it-was-and-is.

Keohane, Robert O. 2006. "The Contingent Legitimacy of Multilateralism." In *Multilateralism Under Challenge*, edited by Edward Newman, Ramesh Thakur and John Tirman. Tokyo: United Nations University Press.

Lango, John. 2014. *The Ethics of Armed Conflict: A Cosmopolitan Just War Theory*. Edinburgh: Edinburgh University Press.

Lee, Steven. 2012. *Ethics and War*. Cambridge: Cambridge University Press.

Matlary, Ian. 2004. "The Legitimacy of Military Intervention: How Important is a UN Mandate?" *Journal of Military Ethics* 3 (2): 129–41.

McMahan, Jeff. 2009. *Killing in War*. Oxford: Clarendon Press.

Parry, Jonathan. 2015. "Just War, Legitimate Authority, and Irregular Belligerency." *Philosophia* 43: 175–96.

Pattison, James. 2008. "Legitimacy and Humanitarian Intervention: Who should intervene?" *The International Journal of Human Rights* 12 (3): 395–413.

Recchia, Stefano. 2016. "Authorising Humanitarian Intervention: a Five-Point Defence of Existing Multilateral Procedures." *Review of International Studies* 43 (1): 50–72.

Reichberg, Gregory. 2005. "Legitimate Authority, Just Cause, and the Decision to Invade Iraq." In *Ethics, Law and Society*, vol. I, edited by Jennifer Gunning and Søren Holm, 243–47. Ashgate: Aldershot.

Reichberg, Gregory. 2012. "Legitimate Authority: Aquinas's First Requirement of a Just War." *Thomist: A Speculative Quarterly Review* 76 (3): 337–69.

Rocheleau, Jordy. 2007. "Preventive War and Lawful Constraints on the Use of Force: An Argument Against International Vigilantism." In *Rethinking the Just War Tradition*, edited by Michael Brough, John Lango, and Harry van der Linden. Albany: SUNY Press.

Schwenkenbecher, Anne. (2013). "Rethinking Legitimate Authority." In *Routledge Handbook of Ethics and War*, edited by Fritz. Allhoff, Nicholas Evans, and Adam Henschke, 161–70. New York: Routledge.

Smith, M. B. E. 1996. "The Duty to Obey the Law." In *A Companion to Philosophy of Law and Legal Theory*, edited by Dennis Patterson, 465–74. Malden, Mass: Blackwell Publishers.

Steinhoff, Uwe. 2009. "What is War and Can a Lone Individual Wage One?" *International Journal of Applied Philosophy* 21 (1): 133–50.

Steinhoff, Uwe. 2019. "Doing Away with 'Legitimate Authority'." *Journal of Military Ethics* 18 (4): 314–32.

Syse, Henrik, and Helene Ingierd. 2005. "What Constitutes a Legitimate Authority?" *Social Alternatives* 24 (3): 11–16.

Thompson, Janna. 2005. "Terrorism, Morality and Right Authority." In *Ethics of Terrorism & Counter-Terrorism*, edited by Georg Meggle, 151–60. Frankfurt: Ontos Verlag.

United Nations. 1945. "The Charter of the United Nations." Accessed November 25, 2020. https://www.un.org/en/sections/un-charter/un-charter-full-text/.

Walzer, Michael. 1977. *Just and Unjust Wars*. New York: Basic Books.

Wellman, Vincent. 1996. "Authority of Law." In *A Companion to Philosophy of Law and Legal Theory*, edited by Dennis Patterson, 583–90. Malden, Mass: Blackwell Publishers.

Williams, Rowan, and George Weigel. 2004. "War and Statecraft: An Exchange." *First Things* March 2004. Accessed June 30, 2016. http://www.firstthings.com/article/2004/03/war-amp-statecraft.

9 Conclusion: applying non-interventionist *jus ad bellum*

9.1 Introduction

Throughout this work, I have argued against moves in just war theory that justify recourse to force too easily. I have defended a non-interventionist view with a high bar of just cause and a requirement of appropriate, presumptively international, authorization. I have criticized the US invasion of Iraq as unjust, both the Iraq and Afghan interventions as continuing too long, and the intervention in Libya as excessive. However, my case for restricting the use of force raises important questions about how to respond to the injustices, suffering, and threats around the world today. What should be done about the civil wars and humanitarian catastrophes in Syria, Yemen, South Sudan, The Democratic Republic of the Congo, and the Central African Republic? How should the world respond to the proliferation of nuclear weapons, including their development or possession by states or sub-state groups thought untrustworthy? What should be done about smaller-scale pockets of terrorists and human rights violators?

These crises challenge anyone theorizing about justice in war to show that our views are practically applicable such that they can and should guide action in difficult situations. Moreover, many would say that these situations call for, if anything, more ready intervention and cast doubt on the anti-interventionist view I have defended. I face the challenge of showing that my view does not implausibly recommend harmful inaction in the face of widespread injustice, human rights violations, suffering, and security threats.

9.2 Considering intervention in Syria and other humanitarian crises and civil wars

Syria has been the most pressing case for humanitarian intervention over the past ten years. The world community's inaction in this situation presents a challenge to and potential indictment of my case for anti-interventionism. Since the civil war in response to government repression began in Syria in 2011, over 400,000 people have been killed, and over five million refugees had fled the state as of 2018 (Independent International

DOI: 10.4324/9781003105381-9

Commission of Inquiry on the Syrian Arab Republic 2020). Although some of the casualties are combatants targeted justifiably in the civil war, many involve war crimes and human rights violations by the Assad government, which has indiscriminately bombed rebel-held areas and exacted punitive revenge on political dissidents. Indeed, the government's violent crackdown on protesters led to the armed civil war. Rebel groups, too, have committed human rights violations, particularly since Isis and Al Qaeda affiliated groups emerged to vie with the government and other rebels for the control of territory (Yassin-Kassab and Al-Shami 2018).

Despite the world community's embrace of the Responsibility to Protect and promise that "never again" would it remain inactive in the face of genocide, there has been a lack of international intervention to stop the civil war, either by a regime change mission or a restricted peacekeeping one. The US and other NATO powers appear reluctant to engage in another large-scale intervention following the prolonged, difficult, and not entirely successful occupations of Afghanistan and Iraq. Adding to the reluctance, in this case, has been a stated concern that intervening to overthrow the Assad regime might enable either a take-over by Isis or other human rights abusers, which could be as bad or worse than Assad as well as a potential sponsor of global terrorism.

Of course, the fact that states do not want to do it does not mean that it is not justified. Indeed unlike the Iraq War, it is difficult to avoid the conclusion that there has been a just cause for intervention in Syria. The magnitude of the human rights crisis combined with the primary responsibility of the Assad regime for the injustice provides a just cause for intervention to stop these humanitarian abuses. Such an intervention would almost certainly satisfy the restrictive mass atrocity criterion that I have defended. However, to be justified, a war must also meet the other *ad bellum* criteria. To satisfy last resort, a good faith attempt to negotiate peace is necessary to satisfy. Such negotiations should not be conditioned on demand for Assad to leave power. However, if Assad is unwilling to negotiate a peace that foreswears a brutal crackdown, then last resort would be satisfied.

The most challenging criteria to apply in this case are those of proportionality and likely success. The worry is that intervention could fail to establish a rights-respecting state while escalating and prolonging the conflict, possibly creating lasting anarchy, an emboldened Isis and Al Qaeda, and at great cost to interveners. I take it that it has not been clear that the pro-democratic rebels could win the civil war and govern the Syrian territory even with international assistance. If it is possible at all, this would require a large-scale ground intervention and significant post-war occupation. Russian support of Assad makes the government harder to topple and intervention prone to escalate conflict. The wars in Iraq and Afghanistan have shown that while US and NATO military might can readily overthrow governments, "winning the peace" and establishing a rights-respecting regime are more challenging. I take it that it is an open question whether a

Syrian intervention would lead to a net gain in human rights and well-being. The toll of the ongoing civil war and rights abuses has been so high that the benefits of overthrowing the Assad regime conceivably outweigh even the tremendous foreseeable costs. The calculation depends upon a factual assessment of military effectiveness, prospects for political leadership and organization, the popular reaction to events, the intentions of actors, and so forth that is immensely complicated. I do not have sufficient information to determine whether it is proportionate and would defer to others authorized to make the decision based upon accurate intelligence, broad consultation, and experienced judgment. Such an intervention could be (or perhaps could have been, if its window has passed, as seems likely) judged beneficial and worthwhile. However, the uncertainties noted here imply that a refusal to conduct a major intervention, especially one aimed at regime change, has not been unreasonable.

The other barrier to justifying intervention in Syria is that of proper authority. Typically, as I argued in Chapter Eight, interventions should be authorized by the UNSC, which has not happened yet in this case. As is well known, Russia, which has been providing military assistance to the Assad government, is certain to veto a proposed intervention. For self-interested reasons as well as the moral reasons just discussed, the Western democracies themselves appear to be reluctant to propose an intervention. In terms of the procedural authorization requirement, before intervening, willing states should make a formal proposal and defend it publicly against Russian and other objections. As I argued in the last chapter, the decision could be referred to an alternative authority if the failure to authorize at the highest level is clearly flawed. A representative regional body, in this case, the Arab League, could be appealed to. A self-authorizing unilateral action would be unjustified without very good evidence of further specious negative decisions at these levels. As I argued above, given the uncertain proportionality and success of intervention, parties would probably be obligated to defer to these international authorities.

None of this is to say that the world community should not do more to attempt to alleviate the human rights crisis and influence a relatively just conclusion, even without military intervention. Foreign states might use their power to establish safe zones inside or outside Syria and use diplomatic pressure to stop the violence even if Assad remains in power.

Notably, the US and allies intervened with a small force in Syria to combat Isis and Al Qaeda forces there on the grounds that this was required to defend against international attacks from those terror groups. They defended their infringement of Syrian sovereignty on the basis that the state was unwilling or unable to stop these attacks. Although not consented to by the Syrian government, the intervention was coordinated with local Kurdish and other pro-democratic groups. It could be objected that this intervention intended to undermine Syrian control and thus contributed to the very inability that was its purported justification (Bridgeman 2018). After the defeat of Isis forces, the US subsequently left a stabilizing force to help protect the zone controlled by the Kurdish forces. In 2019 the Trump administration abruptly withdrew forces from that zone, clearing the way for an incursion and domination by Turkish and Russian forces (Mogelson 2020). Continued minimal peacekeeping, consented to by local parties, with the goal of providing stability and preventing atrocities would have been justified and arguably was morally obligatory. An earlier, larger-scale peacekeeping intervention to protect the territory held by rebel groups from government and terrorist atrocities may have been the most likely measure to proportionately succeed and was more likely justified than a regime change operation.

Interventionists will argue that anti-interventionism such as mine has prevented action that could have ended the crisis in Syria. I reject this for three reasons. First, as I just demonstrated, despite my general resistance to the resort to war, my view allows for intervention in the worst cases, including conditionally in the present and recent Syrian situation. Second, as I argued above, it isn't clear that intervention would successfully end the crisis. Insofar as objections surrounding the likely success, proportionality, and authorization of the war have played a role in checking intervention, this is appropriate. Third, to the extent that intervention has been unjustly ruled out, it isn't due to pacifist and restrictive just war theories. The Western powers have been held back not so much by moral and legal arguments as by self-interested reluctance to commit new lives and resources after the costly wars in Afghanistan and Iraq. The recent US pull-back from its more minimal peacekeeping mission resulted from the Trump administration's own nationalistic commitments.

Unfortunately, Syria is not the only site of large and seemingly intractable human rights abuses. Human Rights Watch's 2020 Report discusses human rights abuses in at least 95 states over the course of 2019. Although I cannot give an exhaustive list, as of this writing, there are several other places that could be argued to satisfy just cause regarding the magnitude of the humanitarian crisis and culpability of local actors. These include Yemen, the Central African Republic, the Democratic Republic of the Congo, and South Sudan (Human Rights Watch 2020). Each has had casualties in excess of a hundred thousand and is characterized by human rights abuses, and appears unlikely to be readily resolved without intervention.

I am not qualified to assess the viability of diplomatic solutions to reach a peace agreement, including the cessation of the assaults on basic human rights. The exhaustion of the possibility of such negotiations is necessary to meet last resort. In Yemen, Saudi Arabia, using arms from the United States, has supported the rights-abusing government. Withdrawal of support and pressure to negotiate could end that conflict (Basri 2018). The Trump administration has been a willing supporter of human rights-violating regimes, including the Duterte government in the Philippines, Modi's nationalism in India, China's abuse of the Huiguhr Muslims, and the actions of Putin's Russia in Ukraine, as well as the Saudis. The first thing for global powers to do is to withdraw support from human rights abusers. Secondly, before undertaking risky, costly, and destructive new armed interventions, more serious effort should be put into creative, diplomatic humanitarian engagement, backed by economic and political pressure.

If last resort is met in some of the humanitarian disaster areas, each crisis poses difficulties regarding the proportionality and likely success of a potential intervention along the lines discussed above regarding Syria. Again, I cannot evaluate the prospects for a net improvement in human rights through military intervention, which hinge on empirical facts and strategic projections beyond the scope of just war theory. Finally, in each case, legitimate procedural authorization would be necessary to establish the representativeness and reasonableness of the intervention.

An additional problem is that intervention might be justified and even obligatory in two or more of these situations at once. This would almost certainly be more than the military powers in the international community would agree to intervene in. Nor could they probably effectively, and thus proportionately, intervene in multiple cases even if they could in one of them. The upshot is that intervention will necessarily be selective. This will result in seeming or real arbitrariness and inconsistency, a barrier to legitimizing the interventions and securing the support necessary for their effectiveness. A degree of inconsistency does not refute a particular intervention as unjust. However, the decision about where to intervene will be more defensible to the extent that it is based upon the direness of the humanitarian crisis and prospects for improving it rather than the self-interested economic and geo-political motives the interveners may have. The possible need for other interventions is a reason to avoid intervention in cases that are not clearly justified, in accord with the thesis of this work.

9.3 Rogue states, nuclear weapons, and preventive war

Humanitarian catastrophes are not the only situations in which military force is being considered. Another which receives much attention is the possible threat posed by so-called "rogue states" – states which have been aggressive to neighbors and hostile to human rights – which are developing nuclear weapons or other weapons of similar mass destruction. Primary discussion in the United States is about preventive strikes against Iran and North Korea. The case for intervention is that it is too dangerous to wait until such states launch an attack or have the clear ability and intent to attack as required by the classic legalist preemptive war doctrine. The rogue state, it is held, can and should be disarmed or overthrown before its threat becomes imminent.

Much has been written on preventive strikes (Shue and Rodin 2007). Most agree that they are morally troubling. Some follow Walzer (1977) in rejecting the prevention of gathering threats while accepting the preemption of imminent ones. Others have argued that preventive strikes can be justified in rare situations when the threat is large and relatively clear. My approach wherein war must pass principled tests, including a consideration of the consequences and general acceptability of norms, supports a restrictive view of preventive war. Preventive strikes generally lack just cause. A party that has not begun to attack has not committed an injustice making it liable to attack in turn. By striking first, a preventer commits precisely the sort of unjust aggressive act that it sought to prevent.

A preventive war almost inevitably fails to be a last resort. When no attack is imminent, there is time to prevent the gathering threat through negotiation or strategic manoeuvers short of war. In the cases of Iran and North Korea, negotiations may help to stop or limit their weapons production, or to the extent that they have them, make them less likely to use them by reducing tension and giving them a stake in international cooperation. Economic carrots and sticks may dissuade investment in weapons as opposed to other productive opportunities. The greater the shared commitment of world and regional actors in joining to prevent the spread of weapons, the more effective such sanctions, and diplomatic opportunities can be. There is also reason to think that, even if so-called rogue states acquire nuclear or biological weapons, they can be deterred by the threat of decisive response, a strategy that has worked to prevent nuclear war between the US, Russia, and China for 70 years (Pollack 2013).

In terms of consequences, a preventive strike replaces the possibility of a future war with a certain one in the present. Acceptance of preventive war would lead to the multiplication of wars, many of which were avoidable. An attempt to overthrow the North Korean or Iranian state would be extremely costly and lead to prolonged resistance and instability. Moreover, to the extent that the states possess deadly weapons, the attack would trigger their use. An attack on North Korea, in particular, would likely provoke retaliation against South Korea or Japan, with massive casualties. The benefits of preventing an uncertain distant attack are not likely to be proportionate to the harm caused. That any particular rogue state would attack if not attacked in the foreseeable future appears quite unlikely, given the threat of a devastating counterattack. Iran was willing to set aside its pursuit of a nuclear weapons program in order to have economic sanctions lifted in the 2015 multilateral treaty but has begun its pursuit again in retaliation for the Trump administration's withdrawal from the agreement (Sanger and Fassihi 2020). North Korea has some nuclear weapons but does not appear to be a threat to attack the United States, although attacks on regional adversaries are real threats. I take it that there is no evidence of a plan by either North Korea or Iran to launch a first strike against any state. Their desire to achieve nuclear weapons is likely aimed foremost at deterrence, much as other nuclear powers justify their own possession. Deterrent possession of nuclear weapons by North Korea and Iran would serve to protect deeply flawed regimes. However, this situation is not such a crisis that it warrants war to avoid it.

As a general norm, a policy of preventive strikes has even more harmful consequences by encouraging conflicting parties to strike early. To the extent that each is aware of a norm embracing preventive strikes, its reason to fear its rival grows with the acceptance of preventive war, creating a spiral of mutual incentive to strike first underwritten by moral cover. By contrast, a bar on preventive war serves to prevent this cycle of incitement. Randall Dipert reports that game-theoretical modeling has actually shown that prevention is an effective strategy such that its "rare and careful use... is actually likely to make the world more safe" (2006, 49). However, notably, the success of his model assumes that preventive attackers are correct 90% of the time about their adversary's future attack. Given the fundamental uncertainty about others' intentions and future actions, the propensity for fear and mixed motives to drive decisions about anticipatory strikes, and the history of misguided preventive wars in Vietnam and Iraq, this is much too optimistic. I suggest that due to epistemic limitations, preventive attacks are unlikely to prevent a war more than 50% of the time. Dipert might counter that if his recommendation were followed, only wars that really have a 90% accuracy would be undertaken. However, preventive war against gathering threats cannot realistically have that level of confidence; parties' judgment that they do would almost certainly be mistaken. The tendency to overestimate the justice of one's cause is especially likely in such uncertain conditions. Indeed Dipert does not argue that there is any historical case of a clearly justified preventive war according to his epistemic threshold but suggests that advancements in intelligence along with the threat of dangerous weapons proliferation will make preventive war justifiable in the future (41–2). However, I doubt that intelligence will be able to know with near certainty *intent*, beyond mere capacity, to use force. The only cases in which a forthcoming attack can be known with 90% certainty are those which are traditionally labeled 'preemptive,' in which the threatened attack is imminent, and the enemy has taken steps to initiate it. Proponents of preventive war doubt that there is a reason to draw a fixed boundary at imminence and certainty and thus argue for a continuum of justifiability encompassing some prevention of more distant threats. However, without being precise, the restriction to preemption is intelligible and has a moral and practical basis. Preemption is already an exception to the norm of not striking first. To blur the limits of the exception by accepting the prevention of distant, gathering threats is dangerous. The acceptance of preventive war as a just cause would inevitably lead to excessive warfare, making it the sort of cause that must be rejected, per my arguments throughout this text.

Dipert (2006, 49) and other apologists argue that a policy of preventive war would beneficially deter dangerous weapons development, which would be encouraged by a policy of non-prevention. Thus a policy of preventive war might stop threats and avoid the need for actual conflict. Yet, deterrence presupposes that parties will know they will likely be attacked if they pursue nuclear weapons and not attacked if they do not. Since it is difficult to know whether weapons development is occurring and impossible to verify that it is not occurring, a state may be as likely to be preventively attacked whether or not it develops such weapons as Iraq found. Thus preventive war is apt to encourage the development of weapons that are rogues states' best chance of deterring attacks. Indeed the preventive attack on Iraq appears to have incited rather than deterred the pursuit of nuclear programs in North Korea and Iran. Like deontological arguments, rule utilitarian arguments tell against preventive war.

The above arguments apply centrally to any preventive conflict which would amount to war. Not only attempted regime change but also any bombing or other attack which would result in significant loss of life constitutes war from a moral standpoint. Such acts are presumptively wrong and require the high threshold of just cause and last resort, which I have argued for throughout the text.

However, targeted strikes which are planned to avoid casualties, particularly collateral damage to civilians, are not wars. Such actions may be undertaken for causes below war's aggression or atrocity threshold. Paradigmatic of such action is the reported use of the Stuxnet virus by the US and Israel to set back Iran's nuclear weapons development (Lucas 2013). By avoiding casualties, destruction of infrastructure, or visible military damage, such strikes are not acts of aggression. Because the development of weapons of mass destruction is dangerous and destabilizing – and generally illegal under non-proliferation agreements – there could be legitimate norm enforcement actions against such programs. Nor is it likely to trigger a large kinetic counter-strike. Indeed Iran did not protest the attack, probably in large part because of the illicit nature of its program. A repeat of such an attack would be justifiable against North Korea (or against Iran again) and presumably has been attempted without as much success.

A less innocuous option is a targeted kinetic strike on nuclear weapons technology in states such as Iran or North Korea. With precision targeting, the US and allies might be able to destroy weapons capabilities with minimal casualties to armed forces as well as civilians. Examples of such an attack include Israel's 1991 destruction of Iraq's Osirak reactor and a 2007 repeat on a Syrian reactor (Gross 2018). These attacks successfully curtailed the weapons programs of the two states with few casualties and no large-scale retaliation. However, because they are likely to have casualties, including risks to non-combatants, and constitute a visible infringement on state sovereignty, such attacks are harder to justify than cyber-attacks. To qualify as measures short of war, they would have to be planned to avoid casualties and enforce international norms. As law enforcement measures, they require international authorization. It is the lack of such authorization that was the problematic feature of the Israeli attacks above. Such measures should be proposed to and deliberated upon by the UNSC. As I argued in Chapter Eight and elsewhere in the context of discussing preventive war (Rocheleau 2007), these measures provide deliberative testing, show respect for international law and its security system, and represent the interests and perspectives of the world community. As with humanitarian intervention, if the UNSC refused to authorize preventive strikes due to an indefensible veto by a biased party, a regional body could be petitioned to authorize the measure. Under strict targeting limits and legitimate procedures, preventive strikes may have a circumscribed justification.[1]

9.4 Targeted strikes on individual terrorists

The above issues share a focus on whether to resort to force against states, either to stop their harm to their own citizens or prevent their international aggression. However, the most commonly considered uses of force today are strikes against non-state actors. Terrorist groups which intentionally target civilians are typical examples of this threat. However, other groups like armed gangs or pirates might be thought to pose a sufficient threat to human rights to justify defensive military action. Combatting such groups does not require as large and destabilizing a use of force as regime changes or even smaller interventions against state governments. Since 2002 the US has been conducting targeted killing campaigns in places including Pakistan, Yemen, Somalia, Libya, and Syria, against individuals thought to pose a threat or connected to groups that pose a threat (Gusterson 2015). Many of these have been done by drones, though occasionally, paradigmatically in the Bin Laden raid, ground forces are used. These raids have been conducted largely in secrecy, without a declaration of war. Their covert nature, along with the comparatively minor force used in each individual strike or raid, has kept much press and public attention from being directed to them. However, collectively, these have involved significant destruction (Gusterson). Such death and destruction need justification and, whether or not we call these conflicts "wars," just war theory must address their conduct if it is to be relevant to the moral evaluation of the use of armed force today.[2]

Upon analysis of just recourse, these cases typically fall short of the just cause threshold, which I have argued is required to justify war. The individual terrorists or groups frequently do not pose a threat to a foreign state amounting to aggression and are not such large rights violators that they warrant humanitarian intervention. I am not able to comment systematically on the extent to which each attack – much less those contemplated – is a response to a significant human rights violation or an act of war. Presumably, in many cases, the targets do pose significant threats to either foreign or domestic citizens. Arguably, targeted strikes, which are relatively small-scale and aimed discriminately at the target who is liable to attack by planning human rights violations, could be justified at a lower threshold than that for justifying war between states.

However, as I argued in Chapter Five, to have ready recourse to such strikes is problematic. Typically, even targeted strikes involve force of a magnitude that results in bystanders casualties. Although the US government asserts minimal civilian casualties, investigating journalists have found relatively high numbers (Gusterson 2015). That they kill innocents implies that a more rigorous threshold for such force is required, as is typical in justifying war as opposed to police measures. Moreover, while a single targeted strike may not create large harm, the collective attacks entailed by an open-ended targeted strike policy against a group cause destruction on the level of war. This argues for applying the just cause threshold for war to evaluate such policies, a test they generally fail.

An additional consideration is that even targeted strikes, when done across international boundaries without authorization by the state where the strike is conducted, violate territorial integrity and state sovereignty. Even if approved by the state, strikes involve the foreign party in a political struggle. The strikes cause conflict between nations and implicate international law relating to the relative rights of parties in the world community. While the strikes do not get a great deal of attention in the US press, these killings from the sky by a powerful foreign power are significant causes for grievance among populations in the areas targeted (Gusterson). Targeted strikes conducted across borders without regard for local sovereignty or international approval mean a form of imperialism in which developing states and their citizens have fewer rights than do their counterparts in wealthy, militarily powerful states.

It could be argued that the US strikes are part of an ongoing "war on terror." After the 2001 terror attacks, it was probably correct for a time to say that the US was at war with Al Qaeda. While this wouldn't justify immediately striking Al Qaeda members anywhere, in violation of sovereign states, it can justify striking them if the states refuse to cooperate with procedures to apprehend them, i.e. are "harboring" them. However, now that the main parties involved in the 9/11 attacks have been killed or apprehended and much time has passed since the attacks, there is not a continuing national defense justification for attacking groups for an Al Qaeda affiliation. Moreover, a loose affiliation with Al Qaeda's overarching mission could not justify attacking groups that are not involved in planning or executing attacks against the US.

I take it that many of the drone strikes in recent years have been taken against parties who cannot be said to have been at war with the attacker. It may be alleged that various terrorist groups are planning attacks, although they haven't carried them out yet. This would make strikes against these groups preventive, with the problems discussed in the previous section. Individuals who are not engaged in fighting are presumably not liable to attack. Active preparation, conspiracy to attack, could be argued to make parties responsible for an unjust imminent threat, thus liable to attack on these grounds. However, if an attack is known with significant certainty to be forthcoming, then one could work with local authorities to have the parties arrested (Miller 2013). A general principle permitting targeted strikes at a low just cause threshold is not universalizable due to its troubling consequences

(Waldron 2012). As I argued in Chapter Five, it is obligatory to use a policing rather than military paradigm if possible. To justify uninvited cross-border violence, the just cause threshold of imminent national defense or atrocity prevention is applicable, as is the criterion of last resort. Even armed intervention should seek to arrest rather than kill if possible, a principle that was violated in the bin Laden raid, among other targeted killings.

As military acts, targeted strikes must satisfy the legitimate authority requirement. While the US administration has claimed legitimacy in the authorization of these strikes, the absence of institutional oversight and even public debate is morally dangerous and arguably illegal. Secret cross–border targeted killings, of the sort the US conducts by drone or Special Forces, are particularly problematic in this regard. While undertaken by a national military, under a chain of command leading to the US President, they violate the letter of international law and bypass international decision-making bodies. In cases of immediate national defense or an imminent, large-scale threat to human rights, states are justified in acting. However, the typical case of a preventive strike against individuals to whom evidence points as planning terror attacks is more problematic. There is an argument for acting without petitioning for international approval insofar as there may be time-sensitive evidence with a narrow window of opportunity. Moreover, public announcement of the plan would likely undermine the effectiveness of the strike. Nonetheless, the selection of targets should have international vetting. If it is impossible to act quickly enough with such vetting, then there should be at least *post hoc* accountability. That is, the Security Council could review the justification of the attacks, certifying valid ones and handing out sanctions for the invalid. In terms of *national* accountability, targeted strikes should have not only institutional oversight but also be announced to the public. If surprise is necessary for effectiveness, then the attacks can be publicized after the fact. President Trump's recent rescinding of public notices of civilian casualties in drone strikes (Savage 2019) further erodes accountability, confirming the presumptive unjustifiability of the measures. Public deliberation and accountability are necessary for ensuring the legitimacy and reasonableness of the use of force.

9.5 Conclusion: the practicality of internationalist modified legalism

I have defended a non-interventionist, internationalist modified legalist just war theory. I have made a case that armed intervention and lethal force are difficult to justify and should not be an ordinary tool of politics. The normative argument presupposes that such a non-interventionism could be observed in practice. In this chapter, I have applied the framework to current crises, including cases that many think call for a more interventionist just war theory. I have sought to thereby demonstrate the practicality of my view beyond the intellectual persuasiveness supported in the first eight chapters. The restrictive norms defended in this work are not utopian. They are reflected in current international law and, to a significant extent, observed. Following these rules is almost always feasible for individual parties and has benefits that counterbalance any sacrifice of national interests.

A theory is never self-implementing, even if taken to be intellectually persuasive. However, I take it that just war theory can influence conduct as its concepts reach leaders and become embedded in organizational culture. In a democratic society, the deliberations of the wider public can trickle up informally through the public sphere so that ideas have several ways of influencing conduct. Of course, anti-interventionist just war theory will only be effective if it is taken up in practice by political leaders and institutions. Concomitant to the goal of avoiding war is the need to bolster diplomatic measures. The US, in particular, spends more than fifteen times as much annually on defense than on diplomacy and foreign aid (CBO 2020; Morello 2020). There is great potential to expand diplomacy and the provision of humanitarian relief, promoting human rights without violence. Were such actions and institutions given the resources and celebrated with the esteem accorded to military institutions, alternatives to war could be more effective. There is still a need and potential for the inculcation of what William James called the "moral equivalent of war," collective social projects which create community solidarity without engaging in violence and entering a life and death struggle with a hated enemy. As pacifists have long argued, it is possible to refocus public energy on non-military means of addressing international political problems. One need not be a pacifist to recognize that a significant policy shift of this sort is desirable and feasible.

My proposal – with its emphasis on global legitimacy and observance of the laws of war – requires international cooperation. It will be more successful to the extent that a responsible culture of international justice, including a foundation in just war theory, in particular, guides the world community, particularly global institutions, and most especially the UN. As a proposal that emphasizes appealing and deferring to international institutions as legitimate authorities, it will be more effective to the extent that states take responsibility for cooperating in good faith to maintain the global security system. At the same time, international institutions must mature in their sense of responsible and just decision-making. As global justice depends on reciprocity, any progress or regress will tend to multiply as it ripples internationally. Blaming the UN and international community for inaction tends to be done in bad faith by those who are not seeking to promote justice themselves. To make international institutions pursue justice more effectively, well-meaning parties should attempt to work constructively within the system rather than seeking to undermine its legitimacy. Moral argument and institutional reform are mutually supporting, such that advancement in either can lead to a spiral of normative advance. Although there is no need to hypothesize a Kantian perpetual peace, the logic of the action suggested here is one in which peace and the protection of human rights can be progressively achieved.

Notes

1 I did not advocate for similar international authorization of cyber-attacks for two reasons. As a small attack, it does not as clearly require this as do kinetic attacks. Secondly, approval would thwart the secrecy of a cyber-attack and might make it unusable.
2 My concern here, as throughout the text, is with the recourse to force rather than the use of particular means such as the drone strike. The morality and effects of the means likely to be employed are relevant, however, to the ethics of engaging in a war at all.

References

Basri, Mohamad. 2018. "The US Could End the War in Yemen if it Wanted to." *The Atlantic*. September 30, 2018. https://www.theatlantic.com/international/archive/2018/09/iran-yemen-saudi-arabia/571465/

Bridgeman, Tess. 2018. "When Does the Legal Basis for U.S. Forces in Syria Expire?" *Just Security*. March 14, 2018. https://www.justsecurity.org/53810/legal-basis-u-s-forces-syria-expire/

CBO (Congressional Budget Office). 2020. "Discretionary Spending in 2019." Posted April 15, 2020. https://www.cbo.gov/publication/56326

Dipert, Randall. 2006. "Preventive War and the Epistemological Dimension of the Morality of War." *Journal of Military Ethics* 5 (1): 32–54.

Gross, Judah Ari. 2018. "Ending a Decade of Silence, Israel Confirms it Blew Up Assad's Nuclear Reactor," *The Times of Israel*. March 21, 2018. https://www.timesofisrael.com/ending-a-decade-of-silence-israel-reveals-it-blew-up-assads-nuclear-reactor/#gs.g8jzav

Gusterson, Hugh. 2015. *Drone: Remote Control Warfare*. London: MIT Press.

Human Rights Watch. 2020. *World Report 2020*. New York: Human Rights Watch.

Independent International Commission of Inquiry on the Syrian Arab Republic. 2020. United Nations Human Rights Council. Accessed November 27, 2020. https://www.ohchr.org/en/hrbodies/hrc/iicisyria/pages/independentinternationalcommission.aspx

Lucas, George. 2013. "Jus in Silico: Moral Restrictions on the Use of Cyberwarfare." In *Routledge Handbook of Ethics and War*, edited by Fritz Allhoff, Nicholas Evans, and Adam Henschke, 367–81. New York: Routledge.

Miller, Seumas. 2013. "Just War Theory and Counterterrorism." In *Routledge Handbook of Ethics and War*, edited by Fritz Allhoff, Nicholas Evans, and Adam Henschke, 226–35. New York: Routledge.

Mogelson, Luke. 2020. "America's Abandonment of Syria." The New Yorker. April 20, 2020. https://www.newyorker.com/magazine/2020/04/27/americas-abandonment-of-syria

Morello, Carol. 2020. "Trump Administration Again Proposes Slashing Foreign Aid." *Washington Post*. February 10, 2020. https://www.washingtonpost.com/national-security/trump-administration-again-proposes-slashing-foreign-aid/2020/02/10/2c03af38-4c4c-11ea-bf44-f5043eb3918a_story.html

Pollack, Kenneth. 2013. *Unthinkable: Iran, the Bomb, and American Strategy*. New York: Simon & Schuster.

Rocheleau, Jordy. 2007. "Preventive War and Lawful Constraints on the Use of Force: an Argument Against International Vigilantism." In *Rethinking the Just War Tradition*, edited by Michael Brough, John Lango, and Harry van der Linden, 183–204. Albany: SUNY Press.

Sanger, David and Farnaz Fassihi. 2020. "Iran Accelerates Nuclear Program But Offers a Path Back from Confrontation." *New York Times*. November 18, 2020. https://www.nytimes.com/2020/11/18/world/middleeast/trump-iran-nuclear.html

Savage, Charlie. 2019. "Trump Revokes Obama-Era Rule on Disclosing Civilian Casualties from U.S. Airstrikes Outside War Zones." *New York Times*. March 6, 2019. https://www.nytimes.com/2019/03/06/us/politics/trump-civilian-casualties-rule-revoked.html

Shue, Henry, and David Rodin, eds. 2007. *Preemption: Military Action and Moral Justification*. Oxford: Oxford University Press.

Waldron, Jeremy. 2012. "Justifying Targeted Killing with a Neutral Principle?" In *Targeted Killings*, edited by Claire Finkelstien, Jens Ohlin, and Andrew Altman, 112–33. Oxford: Oxford University Press.

Walzer, Michael. 1977. *Just and Unjust Wars*. New York, NY: Basic Books.

Yassin-Kassab, Robin, and Leila Al-Shami. 2018. *Burning Country: Syrians in Revolution and War*, New Edition. London: Pluto Press.

Index

Page numbers followed by "n" refer to notes.

For Product Safety Concerns and Information please contact our EU
representative GPSR@taylorandfrancis.com
Taylor & Francis Verlag GmbH, Kaufingerstraße 24, 80331 München, Germany

www.ingramcontent.com/pod-product-compliance
Lightning Source LLC
Chambersburg PA
CBHW060251220326
41598CB00027B/4056

9 780367 615680